VIBRATION ANALYSIS AND CONTROL SYSTEM DYNAMICS

ELLIS HORWOOD SERIES IN ENGINEERING SCIENCE

STRENGTH OF MATERIALS
J. M. ALEXANDER, University College of Swansea.
TECHNOLOGY OF ENGINEERING MANUFACTURE
J. M. ALEXANDER, R. C. BREWER, Imperial College of Science and Technology, University of London, J. R. CROOKALL, Cranfield Institute of Technology.
VIBRATION ANALYSIS AND CONTROL SYSTEM DYNAMICS
CHRISTOPHER BEARDS, Imperial College of Science and Technology, University of London.
COMPUTER AIDED DESIGN AND MANUFACTURE
C. B. BESANT, Imperial College of Science and Technology, University of London.
STRUCTURAL DESIGN AND SAFETY
D. I. BLOCKLEY, University of Bristol.
BASIC LUBRICATION THEORY 3rd Edition
ALASTAIR CAMERON, Imperial College of Science and Technology, University of London.
STRUCTURAL MODELLING AND OPTIMIZATION
D. G. CARMICHAEL, University of Western Australia
ADVANCED MECHANICS OF MATERIALS 2nd Edition
Sir HUGH FORD, F.R.S., Imperial College of Science and Technology, University of London and J. M. ALEXANDER, University College of Swansea.
ELASTICITY AND PLASTICITY IN ENGINEERING
Sir HUGH FORD, F.R.S. and R. T. FENNER, Imperial College of Science and Technology, University of London.
INTRODUCTION TO LOADBEARING BRICKWORK
A. W. HENDRY, B. A. SINHA and S. R. DAVIES, University of Edinburgh
ANALYSIS AND DESIGN OF CONNECTIONS BETWEEN STRUCTURAL JOINTS
M. HOLMES and L. H. MARTIN, University of Aston in Birmingham
TECHNIQUES OF FINITE ELEMENTS
BRUCE M. IRONS, University of Calgary, and S. AHMAD, Bangladesh University of Engineering and Technology, Dacca.
FINITE ELEMENT PRIMER
BRUCE IRONS and N. SHRIVE, University of Calgary
PROBABILITY FOR ENGINEERING DECISIONS: A Bayesian Approach
I. J. JORDAAN, University of Calgary
STRUCTURAL DESIGN OF CABLE-SUSPENDED ROOFS
L. KOLLAR, City Planning Office, Budapest and K. SZABO, Budapest Technical University.
CONTROL OF FLUID POWER, 2nd Edition
D. McCLOY, The Northern Ireland Polytechnic and H. R. MARTIN, University of Waterloo, Ontario, Canada.
TUNNELS: Planning, Design, Construction
T. M. MEGAW and JOHN BARTLETT, Mott, Hay and Anderson, International Consulting Engineers
UNSTEADY FLUID FLOW
R. PARKER, University College, Swansea
DYNAMICS OF MECHANICAL SYSTEMS 2nd Edition
J. M. PRENTIS, University of Cambridge.
ENERGY METHODS IN VIBRATION ANALYSIS
T. H. RICHARDS, University of Aston, Birmingham.
ENERGY METHODS IN STRESS ANALYSIS: With an Introduction to Finite Element Techniques
T. H. RICHARDS, University of Aston, Birmingham.
ROBOTICS AND TELECHIRICS
M. W. THRING, Queen Mary College, University of London
STRESS ANALYSIS OF POLYMERS 2nd Edition
J. G. WILLIAMS, Imperial College of Science and Technology, University of London.

VIBRATION ANALYSIS AND CONTROL SYSTEM DYNAMICS

C. F. BEARDS, B.Sc., Ph.D., C.Eng., M.R.Ae.S.
Department of Mechanical Engineering
Imperial College of Science and Technology
University of London

ELLIS HORWOOD LIMITED
Publishers · Chichester

Halsted Press: a division of
JOHN WILEY & SONS
New York · Brisbane · Chichester · Toronto

First published in 1981 by

ELLIS HORWOOD LIMITED

Market Cross House, Cooper Street, Chichester, West Sussex, PO19 1EB, England

The publisher's colophon is reproduced from James Gillison's drawing of the ancient Market Cross, Chichester.

Distributors:

Australia, New Zealand, South-east Asia:
Jacaranda-Wiley Ltd., Jacaranda Press,
JOHN WILEY & SONS INC.,
G.P.O. Box 859, Brisbane, Queensland 40001, Australia

Canada:
JOHN WILEY & SONS CANADA LIMITED
22 Worcester Road, Rexdale, Ontario, Canada.

Europe, Africa:
JOHN WILEY & SONS LIMITED
Baffins Lane, Chichester, West Sussex, England.

North and South America and the rest of the world:
Halsted Press: a division of
JOHN WILEY & SONS
605 Third Avenue, New York, N.Y. 10016, U.S.A.

British Library Cataloguing in Publication Data

Beards, C. F.
Vibration analysis and control system dynamics –
(Ellis Horwood series in engineering science)
1. Vibration
2. Control theory
I. Title
II. 531'.32 QA865

ISBN 0-85312-242-3 (Ellis Horwood Ltd., Publishers – Library Edn.)
ISBN 0-85312-294-6 (Ellis Horwood Ltd., Publishers – Student Edn.)
ISBN 0-470-27255-4 (Halsted Press)

Typeset in Press Roman by Ellis Horwood Ltd.
Printed in Great Britain by R. J. Acford, Chichester.

Table of Contents

Author's Preface

High costs and questionable availability of materials, land and other resources, together with increasingly sophisticated analysis and manufacturing methods, have led to lightweight structures, high energy sources and finely tuned control systems. These trends, which will surely continue to an increasing extent in the future as ever higher performance levels are demanded, have exacerbated noise, vibration and control system problems.

There is therefore a real need for all practising engineers and scientists as well as students, to have a good understanding of dynamic analysis methods for predicting vibration amplitudes, dynamic stresses and noise levels in a structure, and methods for determining control system performance. It is also essential to be able to understand and contribute to published and quoted data in this field.

There is a great benefit to be gained by studying vibration analysis and control system dynamics together and having this information in a single text, because the analyses of the dynamics of control systems and the vibration of elastic systems are closely linked. This is because in many cases the same equations of motion occur in the control system as in the vibrating system, and thus the techniques developed in the analysis of one system can be used in the other and vice versa. Furthermore, the results of the analysis of one system can be applied to another.

This book is intended to give practising engineers and scientists, and students of engineering and science to first degree level, a thorough understanding of the principles involved in the analysis of vibration and control system dynamics, and to provide a sound theoretical basis for further study. A number of worked examples have been included.

C. F. Beards, September 1980

Author's Preface

CHAPTER 1

Introduction

The vibration which occurs in most machines and structures is undesirable, not only because of the unpleasant motions, the noise and the dynamic stresses, which may lead to fatigue and failure of the structure, but also because of the energy losses and the reduction in performance which accompany the vibrations. The literature is heavy with accounts of system failures brought about by excessive vibration of one component or another. Because of the devastating effects which unwanted vibrations can have on machines and structures, vibration analysis should be carried out as an inherent part of the design, when necessary modifications can most easily be made in an effort to eliminate vibration, or to reduce it as much as possible. Modifications made after this stage, to prototypes or production samples, are usually costly, difficult to implement and often unsatisfactory. Current trends towards lightweight structures and high speed machines are making this analysis increasingly important: as these trends continue so will the need for vibration analysis grow.

The demands made on automatic control systems are also increasing. Whatever the duty of the system, from satellite tracking to controlling sheet thickness in a steel rolling mill, there is a continual effort to improve performance whilst making the system cheaper and more compact. Accurate and relevant analysis of control system dynamics is therefore very necessary, in order to determine the effects of proposed modifications on the system response, as well as predicting the response of new system designs.

There are two reasons why we should study vibration analysis and control system dynamics together. Firstly, because we need to consider control in relation to mechanical engineering using mechanical analogies, rather than as a specialised and isolated aspect of electrical engineering, and, secondly, because the basic equations governing the behaviour of vibration and control systems are the same: different emphasis is placed on the different forms of the solution available, but they are all dynamic systems. Each subject benefits from techniques developed in the other.

Dynamic analysis can be carried out most conveniently in the following three stages:

Stage I. Form a mathematical or physical model of the system to be analysed.
Stage II. From the model write the equations of motion.
Stage III. Evaluate the system response to relevant specific excitation.

These stages are now considered in greater detail.

Stage I. The mathematical model

To model any real system a number of simplifying assumptions have to be made. For example, a distributed mass may be considered as a lumped-mass, or the effect of damping in the system may be neglected particularly if only resonant frequencies are sought or a non-linear spring may be considered linear, or certain elements may be neglected altogether if their effect is likely to be small. Furthermore the directions of motion of the mass elements are usually restrained to those of immediate interest to the analyst.

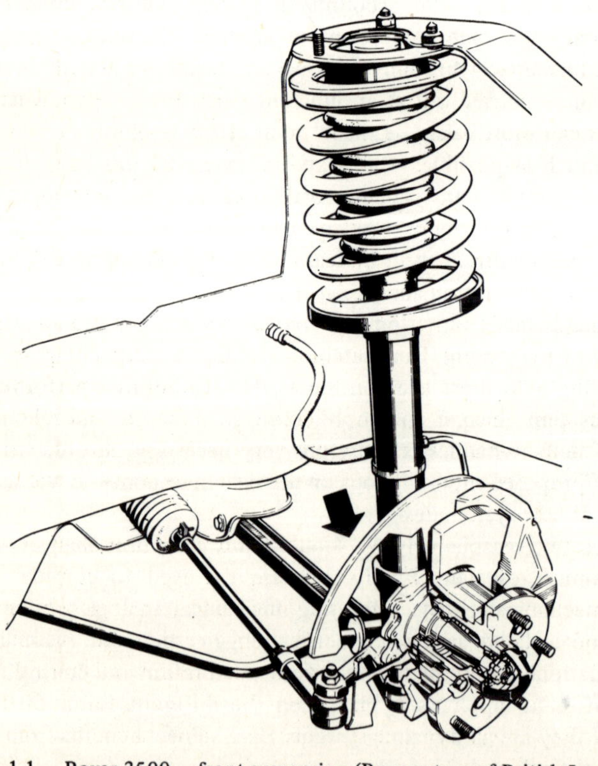

Fig. 1.1 – Rover 3500 – front suspension (By courtesy of British Leyland).

Thus the model is usually a compromise between a simple representation which is easy to analyse but may not be very accurate, and a complicated but realistic model which is difficult to analyse but does give useful results. Consider, for example, the analysis of the vibration of the front wheel of a motor car. Fig. 1.1 shows a typical suspension. Fig. 1.2(a) is a very simple model of this same system, which considers translational motion in a vertical direction only: this model is not going to give much useful information, although it is easy to analyse. The model shown in Fig. 1.2(b) is capable of producing some meaningful

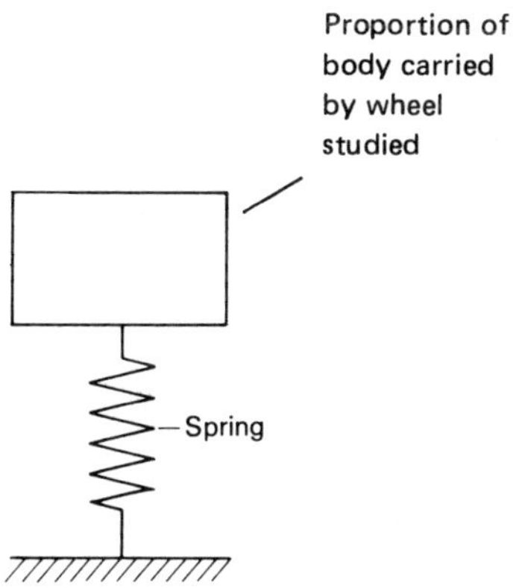

Fig. 1.2(a) – Simplest model – Motion in a vertical direction only can be analysed.

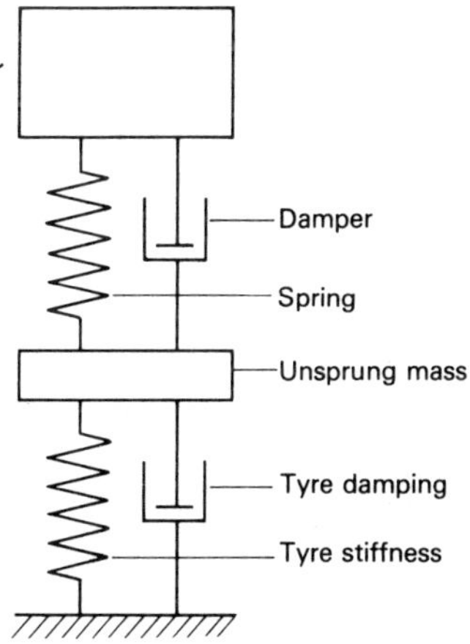

Fig. 1.2(b) – Motion in a vertical direction only can be analysed.

results at the cost of increased labour in the analysis, but the analysis is still confined to motion in a vertical direction only. A more refined model, shown in Fig. 1.2(c), shows the whole car considered, translational and rotational motion of the car body being allowed.

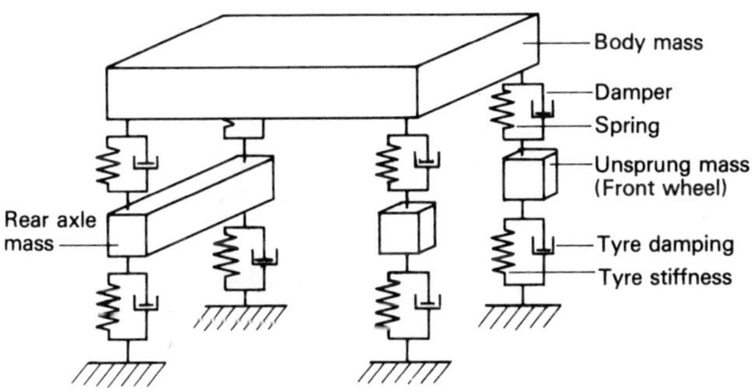

Fig. 1.2(c) – Motion in a vertical direction, roll, and pitch can be analysed.

If the modelling of the car body as a rigid mass is not acceptable, a finite element approach to the analysis may be necessary.

A block diagram model is usually used in the analysis of control systems. For example, a system used for controlling the rotation and position of a turntable about a vertical axis is shown in Fig. 1.3. The turntable can be used for

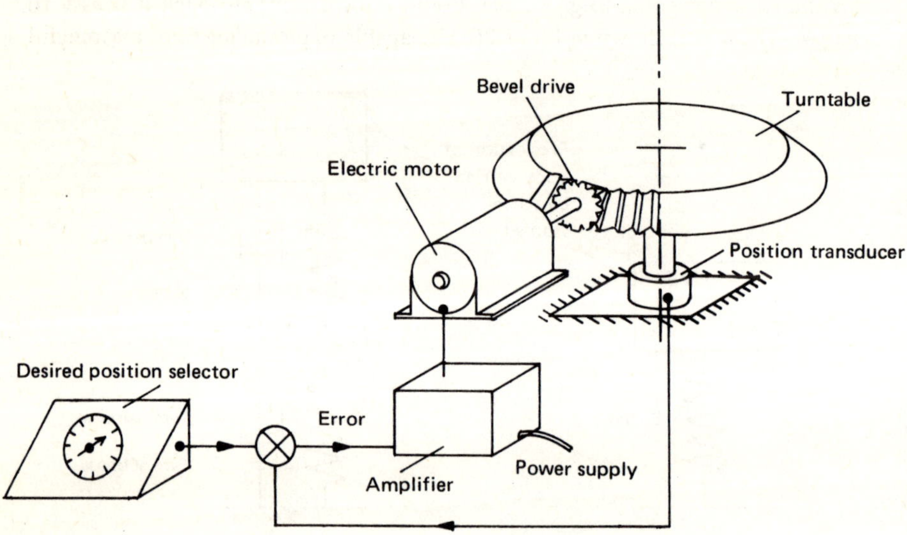

Fig. 1.3 – Turntable position control system.

mounting a telescope or gun, or if it forms part of a machine tool it can be used for mounting a workpiece for machining. Fig. 1.4 shows the block diagram model used in the analysis.

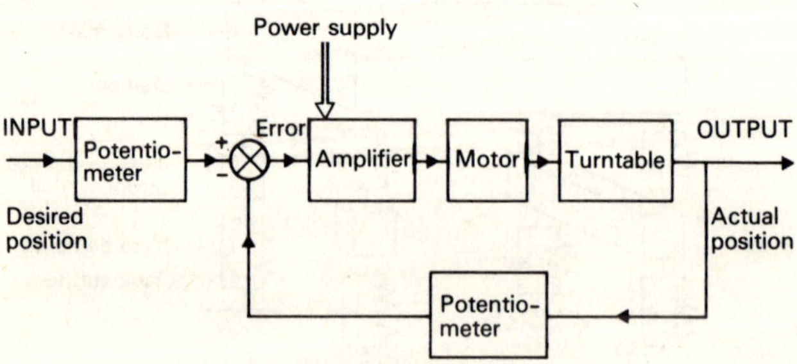

Fig. 1.4 – Turntable position control system: block diagram model.

Model Parameters

Because of the approximate nature of most models, whereby small effects are neglected and the environment is made independent of the system motions, it is usually reasonable to assume constant parameters and linear relationships. This means that the coefficients in the equations of motion are constant and the equations themselves are linear: these are real aids to simplifying the analysis. Distributed masses can often be replaced by lumped mass elements to give ordinary rather than partial differential equations of motion. Usually the numerical value of the parameters can, substantially, be obtained directly, from the system being analysed. However, model system parameters are sometimes difficult to assess, and then an intuitive estimate is required, engineering judgement being of the essence.

It is not easy to create a relevant mathematical model of the system to be analysed, but such a model does have to be produced before Stage II of the analysis can be started. Most of the material in subsequent chapters is presented to make the reader competent to carry out the analyses described in Stages II and III. A full understanding of these methods will be found to be of great help in formulating the mathematical model referred to above in Stage I.

Stage II. The equations of motion

Several methods are available for obtaining the equations of motion from the mathematical model, the choice of method often depending on the particular model and personal preference. For example, analysis of the free-body diagrams drawn for each body of the model usually produces the equations of motion quickly: but it can be advantageous in some cases to use an energy method such as the Lagrange equation.

From the equations of motion the characteristic or frequency equation is obtained, yielding data on the natural frequencies, modes of vibration, general response, and stability.

Stage III. Response to specific excitation

Although Stage II of the analysis gives much useful information on natural frequencies, response and stability, it does not give the actual system response to specific excitations. It is necessary to know the actual response in order to determine such quantities as dynamic stress or noise for a range of system inputs, either force or motion, including harmonic, step and ramp. This is achieved by solving the equations of motion with the excitation function present.

Remember:

Stage I	Model
Stage II	Equations
Stage III	Excitation.

So far we have considered a few examples of how real systems can be modelled and the principles of their analysis. Obviously, if we are to be competent to analyse system models, it is essential to be able to analyse single degree of freedom systems, which are considered in Chapter 2, systems with more than one degree of freedom in Chapter 3, and continuous systems in Chapter 4. Some aspects of automatic control system analysis which require special consideration, particularly their stability and the system frequency response, are discussed in Chapter 5. Chapter 6 contains a number of exercises for the reader to try.

CHAPTER 2

The vibrations of systems having one degree of freedom

A system with one degree of freedom is the simplest case to analyse because only one coordinate is necessary to completely describe the motion of the system. Some real systems can be modelled in this way, either because the excitation of the system is such that the vibration can be described by one coordinate although the system could vibrate in other directions if so excited, or the system really is simple, as for example in the case of a clock pendulum. It should also be noted that a one degree of freedom model of a complicated system can often be constructed where the analysis of a particular mode of vibration is to be carried out. To be able to analyse one degree of freedom systems is therefore an essential ability in vibration analysis.

2.1 FREE UNDAMPED VIBRATION

2.1.1 Translational vibration

In the system shown in Fig. 2.1 a body of mass m is free to move along a fixed horizontal surface. A spring of constant stiffness k which is fixed at one end is attached at the other end to the body. Displacing the body to the right (say) from the equilibrium position causes a spring force to the left (a restoring force). This force gives the body an acceleration to the left. When the body reaches its equilibrium position the spring force is zero, but the body has a velocity which

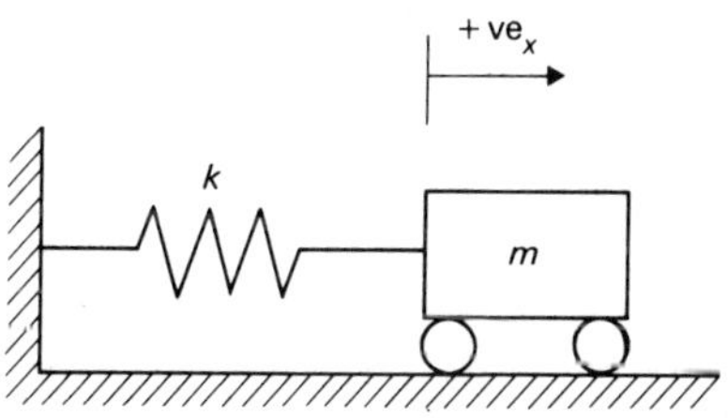

Fig. 2.1

carries it further to the left although it is retarded by the spring force which now acts to the right. When the body is arrested by the spring, the spring force is to the right so that the body moves to the right, past its equilibrium position, and hence reaches its initial displaced position. In practice this position will not quite be reached because damping in the system will have dissipated some of the vibrational energy. However, if the damping is small its effect can be neglected.

If the body is displaced a distance x_0 to the right and released, the free-body diagrams (FBDs) for a general displacement x are as shown in Figs. 2.2(a) and (b).

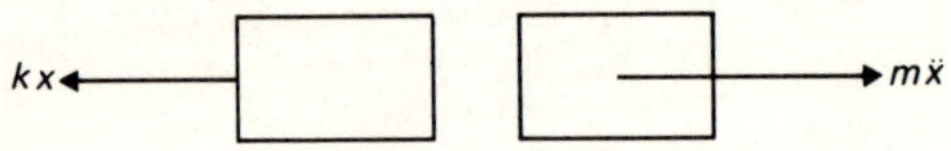

Fig. 2.2(a) – Applied force. (b) – Effective force.

The effective force is always in the direction of positive x. If the body is being retarded $\ddot{x}$ will calculate to be negative. The mass of the body is assumed contant: this is usually so, but not for example in the case of a rocket burning fuel. The spring stiffness k is assumed contant: this is usually so within limits, see section 2.1.3. It is assumed that the mass of the spring is negligible compared to the mass of the body.

From the free-body diagrams the equation of motion for the system is

$$m\ddot{x} = -kx \quad \text{or} \quad \ddot{x} + (k/m)x = 0 \tag{2.1}$$

This will be recognised as the equation for simple harmonic motion.

The solution is $$x = A \cos \omega t + B \sin \omega t \tag{2.2}$$

where A and B are constants which can be found by considering the initial conditions, and ω is the circular frequency of the motion. Substituting (2.2) into (2.1) we get

$$-\omega^2(A \cos \omega t + B \sin \omega t) + (k/m)(A \cos \omega t + B \sin \omega t) = 0$$

since $A \cos \omega t + B \sin \omega t \neq 0$ (otherwise no motion), $\omega = \sqrt{(k/m)}$ rad/s,

and $x = A \cos \sqrt{(k/m)}\, t + B \sin \sqrt{(k/m)}\, t$.

Now $x = x_0$ at $t = 0$ so $x_0 = A$, and $\dot{x} = 0$ at $t = 0$ so $0 = B\sqrt{(k/m)}$

that is, $$x = x_0 \cos \sqrt{(k/m)}\, t \tag{2.3}$$

The system parameters control ω and the type of motion but not the amplitude x_0, which is found from the initial conditions. The mass of the body is important, its weight is not, so that for a given system, ω is independent of the local gravitational field.

$$\text{Frequency, } f = \omega/(2\pi) = 1/(2\pi)\sqrt{(k/m)} \text{ Hz.} \quad (2.4)$$

The motion is as shown in Fig. 2.3.

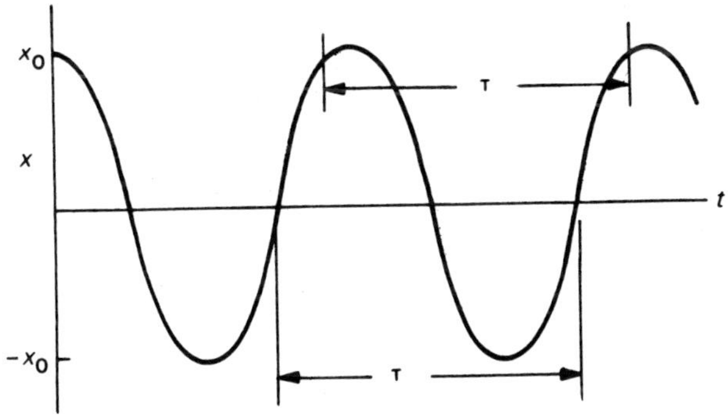

Fig. 2.3

The period, τ, is the time taken for one complete cycle.

$$\tau = 1/f \text{ s} = 2\pi\sqrt{(m/k)} \text{ s} \quad . \quad (2.5)$$

The analysis of the vibration of a body supported to vibrate in the vertical or y direction can be carried out in a similar way to that above. Fig. 2.4 shows the system.

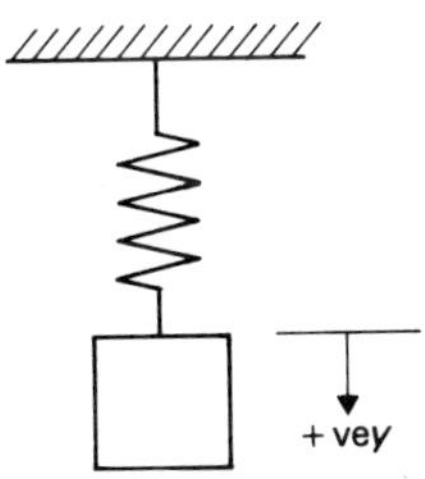

Fig. 2.4

The spring extension δ when the body is fastened to the spring is given by $k\delta = mg$. When the body is given a displacement y_0 and released the FBDs are for a general displacement y, as in Fig. 2.5.

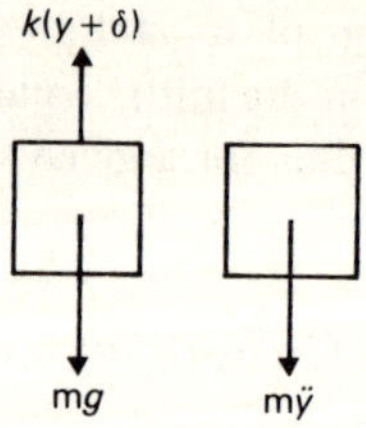

Fig. 2.5(a) – Applied forces. (b) Effective force.

The equation of motion is

$$\begin{aligned} m\ddot{y} &= mg - k(y + \delta) \\ &= mg - ky - k\delta = -ky, \text{ since } mg = k\delta. \end{aligned}$$

that is, $\ddot{y} + (k/m)\,y = 0$. (2.6)

This is similar to equation (2.1), so that the general solution is

$$y = A \cos \sqrt{(k/m)}\, t + B \sin \sqrt{(k/m)}\, t \quad . \tag{2.7}$$

Note that $\sqrt{(k/m)} = \sqrt{(g/\delta)}$. For the initial conditions $y = y_0$ at $t = 0$

$$\text{and } \dot{y} = 0 \text{ at } t = 0,\ y = y_0 \cos \sqrt{(k/m)}\, t \quad , \tag{2.8}$$

that is, the frequency of vibration is the same whether the body vibrates in a horizontal or vertical direction.

Sometimes more than one spring acts in a vibrating system. The spring, which is considered to be an elastic element of constant stiffness, can take many forms in practice; for example, it may be a wire coil, rubber block, beam or air bag. Combined spring units can be replaced in the analysis by a single spring of equivalent stiffness as follows.

1) *Springs connected in series*

The three spring system of Fig. 2.6(a) can be replaced by the equivalent spring of Fig. 2.6(b).

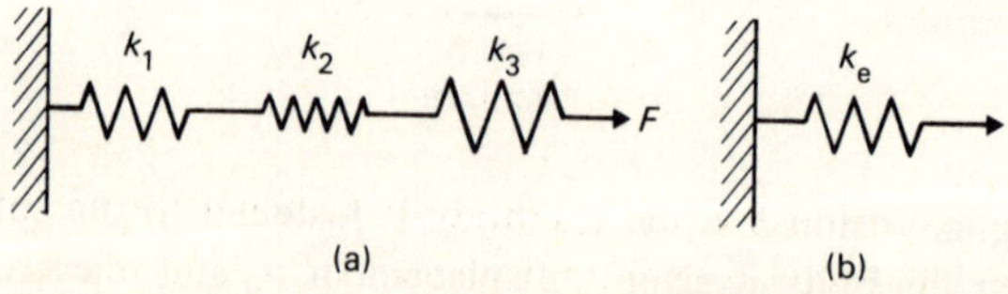

Fig. 2.6

If the deflection δ experienced by applying the force F is to be the same

in both cases, $\delta = F/k_e = F/k_1 + F/k_2 + F/k_3$

that is $1/k_e = \sum_{i=1}^{3} 1/k_i$.

In general, the reciprocal of the equivalent stiffness of springs connected in series obtained by summing the reciprocal of the stiffness of each spring.

2) *Springs connected in parallel*

The three spring system of Fig. 2.7(a) can be replaced by the equivalent spring of Fig. 2.7(b).

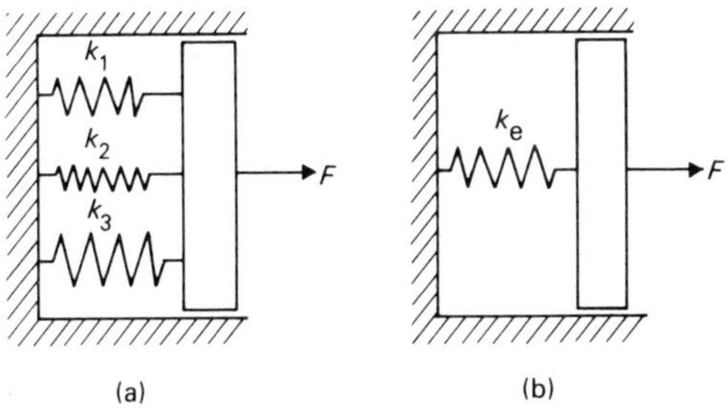

Fig. 2.7

Since the deflection δ must be the same in both cases, the sum of the forces exerted by the springs in parallel must equal the force exerted by the equivalent spring.

Thus $F = k_1\delta + k_2\delta + k_3\delta = k_e\delta$,

that is $k_e = \sum_{i=1}^{3} k_i$

In general, the equivalent stiffness of springs connected in parallel is obtained by summing the stiffness of each spring.

2.1.2 Torsional vibration

Fig. 2.8 shows the model used to study torsional vibration.

A body with mass moment of inertia I about the axis of rotation is fastened to a bar of torsional stiffness k_T. If the body is rotated through an angle θ_o and released, torsional vibration of the body results. The mass moment of inertia of the shaft about the axis of rotation is usually negligible compared with I.

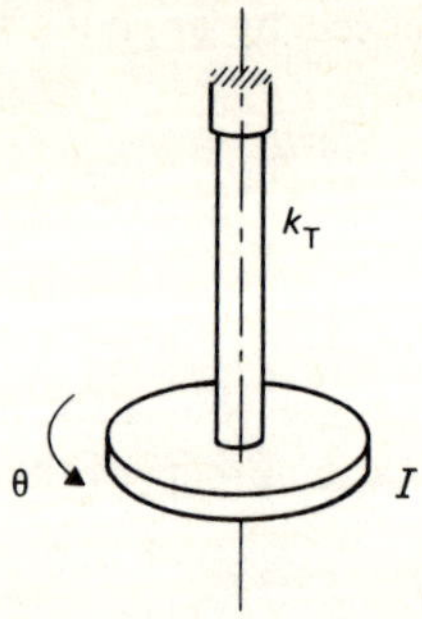

Fig. 2.8

For a general displacement θ the FBDs are as given in Figs. 2.9(a) and (b).

Fig. 2.9(a) – Applied torque. (b) Effective torque.

Hence the equation of motion is

$$I\ddot{\theta} = -k_T\theta \text{ or } \ddot{\theta} + (k_T/I)\theta = 0 \quad .$$

This represents SHM with circular frequency $\omega = \sqrt{(k_T/I)}$ rad/s.
If the spring element is a shaft,

$$k_T = \frac{\text{applied torque}}{\text{angle of twist}} = \frac{GJ}{l}$$

where G = modulus of rigidity for the shaft material,
J = second moment of area about axis of rotation,
and l = the length of shaft.

$$\text{that is, } f = \frac{\omega}{2\pi} = \frac{1}{2\pi}\sqrt{(GJ/Il)} \text{ Hz} \quad .$$

If the shaft does not have a constant diameter, it can be replaced analytically by an equivalent shaft of different length but with the same stiffness and a constant diameter.

For example, a circular section shaft comprising a length l_1 of diameter d_1

and a length l_2 of diameter d_2 can be replaced by a length l_1 of diameter d_1 and a length l of diameter d_1 where, for the same stiffness,

$$(GJ/l)_{\substack{\text{length } l_2 \\ \text{diameter } d_2}} = (GJ/l)_{\substack{\text{length } l \\ \text{diameter } d_1}}$$

that is, for the same shaft material $= d_2{}^4/l_2 = d_1{}^4/l_1$.

Thus the equivalent length l_e of the shaft of constant diameter d_1 is given by

$$l_e = l_1 + (d_1/d_2)^4 l_2 \ .$$

The Torsional Vibration of a Geared System

Consider the system shown in Fig. 2.10. The mass moments of inertia of the

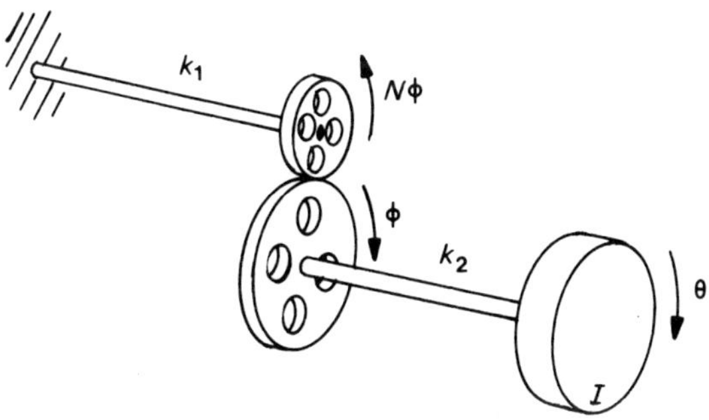

Fig. 2.10

shafts and gears about their axes of rotation are considered negligible. The shafts are supported in bearings which are not shown, and the gear ratio is N:1.

From the FBDs, T_2, the torque in shaft 2 is $T_2 = k_2(\theta - \phi) = -I\ddot{\theta}$ and T_1, the torque in shaft 1, is $T_1 = k_1 N\phi$, since $NT_1 = T_2$, $T_2 = k_1 N^2 \phi$ and $\phi = k_2\theta/(k_2 + k_1 N^2)$.

Thus the equation of motion becomes

$$I\ddot{\theta} + \left(\frac{k_2 k_1 N^2}{k_2 + k_1 N^2}\right) \theta = 0$$

$$\text{and } f = \frac{1}{2\pi}\sqrt{\frac{[(k_2 k_1 N^2)/(k_2 + k_1 N^2)]}{I}} \text{ Hz.}$$

that is, k_{eq}, the equivalent stiffness referred to shaft 2, is $(k_1 k_2 N^2)/(k_1 N^2 + k_2)$, or $1/k_{eq} = 1/k_2 + 1/k_1 N^2$.

2.1.3 Non-linear spring element

Any spring elements have a force-deflection relationship which is linear only over a limited range of deflection. Fig. 2.11 shows a typical characteristic.

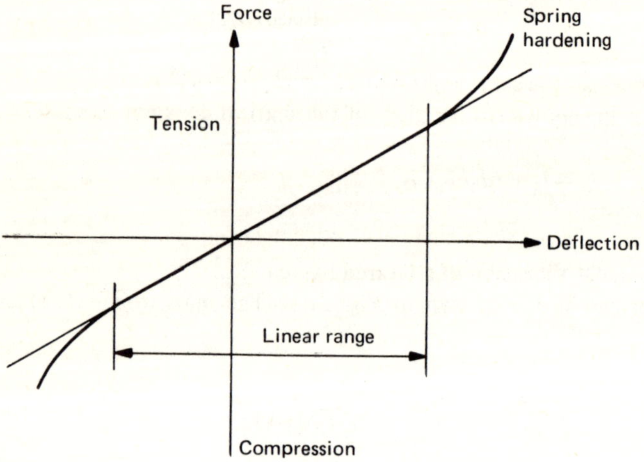

Fig. 2.11

The non-linearities in this characteristic may be caused by physical effects such as the contacting of coils in a compressed coil spring, or by excessively straining the spring material so that yielding occurs. In some systems the spring elements do not act at the same time, as shown in Fig. 2.12(a), or the spring is designed to be non-linear as shown in Figs. 2.12(b) and (c).

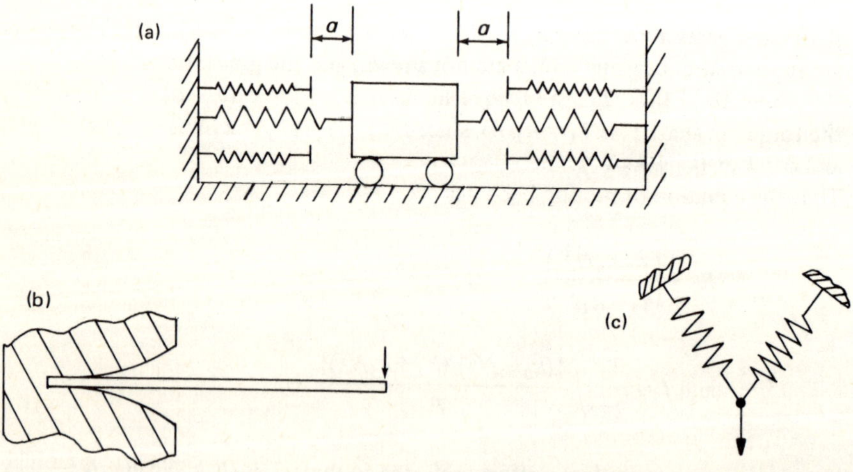

Fig. 2.12

Analysis of the motion of the system shown in Fig. 2.12(a) requires analysing the motion until the half-clearance a is taken up, and then using the displacement and velocity at this point as initial conditions for the ensuing motion when the extra springs are operating. Similar analysis is necessary when the body leaves the influence of the extra springs.

2.1.4 Energy methods for analysis

For undamped free vibration the total energy in the vibrating system is constant throughout the cycle. Thus the maximum potential energy (PE) is equal to the maximum kinetic energy (KE), although these maxima occur at different times. Furthermore

$$\mathrm{d}/\mathrm{d}t\ (\mathrm{KE} + \mathrm{PE}) = 0.$$

Applying this method to the case, already considered, of a body of mass m fastened to a spring of stiffness k, when the body is displaced a distance x from its equilibrium position,

Strain energy (SE) in spring $= \frac{1}{2}\,kx^2$.
Kinetic energy (KE) of body $= \frac{1}{2}\,m\dot{x}^2$.

The SE is the total PE in this case because the weight is equal to the spring force in the equilibrium position.

Thus $\mathrm{d}/\mathrm{d}t\ (\frac{1}{2}\,kx^2 + \frac{1}{2}\,m\dot{x}^2) = 0$,

that is, $kx\dot{x} + m\ddot{x}\dot{x} = 0$,

or $\ddot{x} + (k/m)x = 0$, as before.

This is a very useful method for certain types of problem in which it is difficult to apply Newton's Laws of Motion.

Alternatively, assuming SHM, if $x = x_0 \cos \omega t$

$$\text{maximum SE} = \tfrac{1}{2}\,kx_0^2$$
$$\text{maximum KE} = \tfrac{1}{2}\,m(x_0\omega)^2$$

Thus $\frac{1}{2}\,kx_0^2 = \frac{1}{2}\,mx_0^2\omega^2$ since $\mathrm{SE}_{max} = \mathrm{KE}_{max}$,

or $\omega = \sqrt{(k/m)}$ rad/s.

Example 1

A link AB in a mechanism is of uniform section 0.3 m long. It has a mass of 10 kg, and a concentrated mass of 7 kg is attached at B. The link is hinged at A and is supported in a horizontal position by a spring attached at the mid point of the bar. The stiffness of the spring is 2 kN/m. Find the frequency of small free oscillations of the system. The system is as shown below.

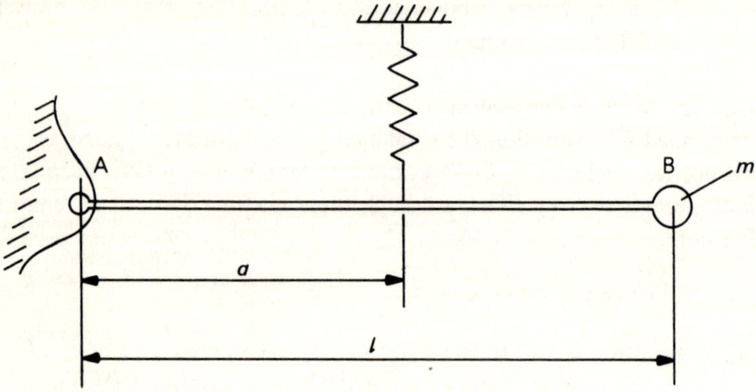

For rotation about A the equation of motion is:

$$I_A\ddot{\theta} = -ka^2\theta,$$

that is, $\ddot{\theta} + (ka^2/I_A)\,\theta = 0$.

This is SHM with frequency $\dfrac{1}{2\pi}\sqrt{(ka^2/I_A)}$ Hz.

In this case $a = 0.15$ m, $l = 0.3$ m, $k = 2000$ N/m,

and $I_A = 7(0.3)^2 + \frac{1}{3}.10.(0.3)^2 = 0.93$ kg m^2.

Hence $f = \dfrac{1}{2\pi}\sqrt{\left(\dfrac{2000.(0.15)^2}{0.93}\right)} = 1.1$ Hz.

Example 2

A small turbo generator has a turbine disc of mass 20 kg and radius of gyration 0.15 m driving an armature of mass 30 kg and radius of gyration 0.1 m through a steel shaft 0.05 m diameter and 0.4 m long. The modulus of rigidity for the shaft steel is 86×10^9 N/m^2. Determine the natural frequency of torsional oscillation of the system, and the position of the node.

The rotors must twist in opposite direction to each other; i.e. along the shaft there is a section of zero twist: this is called a node.

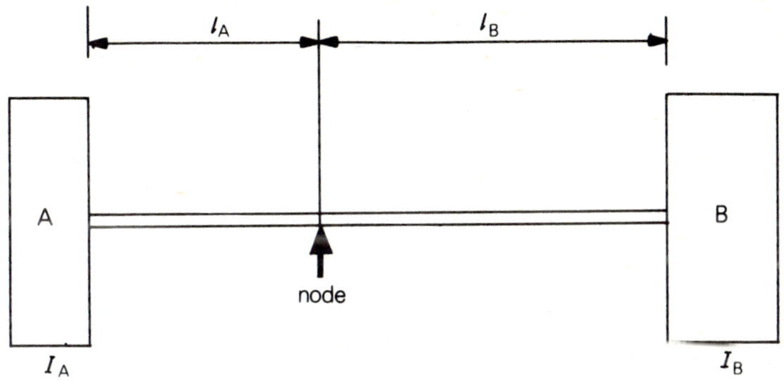

The frequency of oscillation of each rotor is the same, and since there is no twist at the node,

$$f = \frac{1}{2\pi}\sqrt{\left(\frac{GJ}{I_A l_A}\right)} = \frac{1}{2\pi}\sqrt{\left(\frac{GJ}{I_B l_B}\right)},$$

that is, $I_A l_A = I_B l_B$.

Now $I_A = 20(0.15)^2 = 0.45$ kg m^2 and $I_B = 30(0.1)^2 = 0.3$ kg m^2.

Thus $l_A = (0.3/0.45)\, l_B$. Since $l_A + l_B = 0.4$ m

$l_A = 0.16$ m, i.e. the node is 0.16 m from the turbine disc.

Hence $f = \dfrac{1}{2\pi}\sqrt{\left(\dfrac{86.10^9\,\pi(0.05)^4}{0.45\;0.16.32}\right)} = 136$ Hz.

Example 3

A uniform cylinder of mass m is rotated through a small angle θ_o from the equilibrium position and released. Determine the equation of motion and hence obtain the frequency of free vibration. The cylinder rolls without slipping.

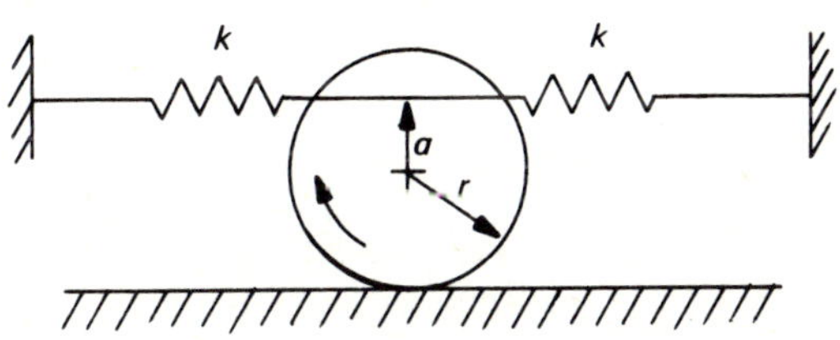

If the axis of the cylinder moves a distance x and turns through an angle θ so that $x = r\theta$,

$$\text{KE} = \tfrac{1}{2} m\dot{x}^2 + \tfrac{1}{2} I\dot{\theta}^2, \text{ where } I = \tfrac{1}{2} mr^2.$$

Hence $\text{KE} = \tfrac{3}{4} mr^2 \dot{\theta}^2$

$$\text{SE} = 2.\tfrac{1}{2}.k[(r+a)\theta]^2 = k(r+a)^2\theta^2.$$

Now energy is conserved, so $(\tfrac{3}{4} mr^2\dot{\theta}^2 + k(r+a)^2\theta^2)$ is constant,

that is, $\mathrm{d}/\mathrm{d}t\,(\tfrac{3}{4} mr^2\dot{\theta}^2 + k(r+a)^2\theta^2) = 0$

or $\tfrac{3}{4} mr^2\, 2\dot{\theta}\ddot{\theta} + k(r+a)^2 2\theta\dot{\theta} = 0$.

Thus the equation of motion is $\ddot{\theta} + \dfrac{k(r+a)^2\theta}{3/4\, mr^2} = 0.$

Hence frequency of free vibration is $\dfrac{1}{2\pi}\sqrt{\left[\dfrac{4k(r+a)^2}{3mr^2}\right]}$ Hz.

Example 4

In the system shown in Fig. 2.12(a) find the period of free vibration of the body if it is displaced a distance x_0 from the equilibrium position and released, when x_0 is greater than the half clearance a. Each spring has a stiffness k, and damping is negligible.

Initially the body is under the action of four springs so that the equation of motion is $m\ddot{x} + 4kx = 0$. The general solution is $x = A\cos\omega t + B\sin\omega t$, but with initial conditions $x = x_o$ at $t = 0$ and $\dot{x} = 0$ at $t = 0$ the solution is $x = x_o\cos\omega t$.

When $x = a$, $\quad a = x_o\cos\omega t_1$,

that is, $t_1 = (1/\omega)\cos^{-1}(a/x_o) = \sqrt{(m/4k)}\cos^{-1}(a/x_o)$.

This is the time taken for the body to move from $x = x_o$ to $x = a$. The velocity of the body at time t_1 is $-x_o\sqrt{(4k/m)}\sin\omega t_1$,

that is, $-(4k/m)\sqrt{(x_o^2 - a^2)}$.

For motion from $x = a$ to $x = 0$, the body moves under the action of two springs only, with the initial conditions $x_o = a$ and $\dot{x} = -(4k/m)\sqrt{(x_o^2 - a^2)}$.

The equation of motion is $m\ddot{x} + 2kx = 0$, the general solution to which is $x = C \cos \Omega t + D \sin \Omega t$, where $\Omega = \sqrt{(2k/m)}$.

Since $x_o = a$ at $t = 0, C = a$

and because $\dot{x} = -(4k/m)\sqrt{(x_o^2 - a^2)}$ at $t = 0, D = \sqrt{(2(x_o^2 - a^2))}$.

That is, $x = a \cos \sqrt{(2k/m)}\, t - \sqrt{(2(x_o^2 - a^2))} \sin \sqrt{(2k/m)} t$.

When $x = 0$, $a \cos \sqrt{(2k/m)}\, t_2 = \sqrt{(2(x_o^2 - a^2))} \sin \sqrt{(2k/m)}\, t_2$,

that is, $t_2 = \sqrt{(m/2k)} \tan^{-1} a/\sqrt{(2(x_o^2 - a^2))}$.

Thus the time for $\frac{1}{4}$ cycle is $t_1 + t_2$, and the time one cycle, i.e. the period of free vibration, is

$$4[\sqrt{(m/4k)} \cos^{-1}(a/x_o) + \sqrt{(m/2k)} \tan^{-1} a/\sqrt{(2(x_o^2 - a^2))}] \text{ s.}$$

2.2 FREE DAMPED VIBRATION

All real systems dissipate energy when they vibrate. This is due to fricitional effects, for example that occurring at the connection between elements, internal friction in deformed members, and windage. It is often difficult to model damping exactly because many mechanisms may be operating in a system. However, each type of damping can be analysed, and in many systems one form of damping dominates so that a reasonably accurate analysis is possible. The most common types of damping are viscous, dry friction, and hysteretic.

2.2.1 Vibration with viscous damping

Viscous damping is a common form of damping which is found in many engineering systems such as instruments and shock absorbers. The viscous damping force is proportional to the first power of the velocity across the damper, and it always opposes the motion, so that the damping force is a linear continuous function of the velocity. Because the analysis of viscous damping leads to the simplest mathematical treatment, analysts sometimes approximate more complex types of damping to the viscous type.

Consider the single degree of freedom model with viscous damping shown in Fig. 2.13.

The only unfamiliar element in this system is the viscous damper with coefficient c. This coefficient is such that the damping force required to move the body with a velocity $\dot{x}$ is $c\dot{x}$.

For motion of the body in the direction shown, the free body diagrams are as in Fig. 2.14(a) and (b).

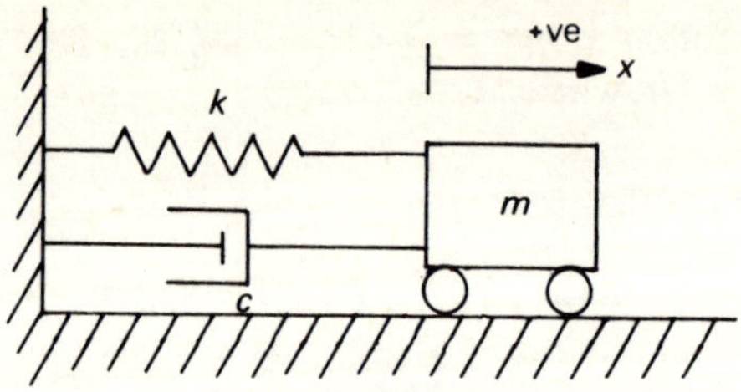

Fig. 2.13

Fig. 2.14(a) – Applied forces. (b) Effective force.

The equation of motion is therefore

$$m\dot{x} + c\dot{x} + kx = 0 \quad . \tag{2.9}$$

This equation of motion pertains to the whole of the cycle: the reader should verify that this is so (Note: displacements to the left of the equilibrium position are negative, and velocities and accelerations from right to left are also negative).

Equation (2.9) is a second order differential equation which can be solved by assuming a solution of the form $x = Xe^{st}$. Substituting this solution into equation (2.9) gives

$$(ms^2 + cs + k)Xe^{st} = 0.$$

Since $Xe^{st} \neq 0$ (no motion), $ms^2 + cs + k = 0$,

that is, $s_{1,2} = -c/2m \pm \sqrt{(c^2 - 4mk)}/2m$.

Thus $x = X_1 e^{s_1 t} + X_2 e^{s_2 t}$.

X_1 and X_2 are arbitrary constants found from the initial conditions. The response evidently depends on whether c is positive or negative and on whether c^2 is greater than, equal to, or less than $4mk$.

The behaviour of the system depends on the numerical value of the radical, so we define **critical damping** c_c as that value of c which makes the radical zero.

Thus $c_c/2m = \sqrt{(k/m)} = \omega$, the undamped natural frequency,

and $c_c = 2\sqrt{(km)} = 2m\omega$.

The actual damping in a system can be specified in terms of c_c by introducing the **damping ratio** ζ.

Thus $\zeta = c/c_c$

and $s_{1,2} = (-\zeta \pm \sqrt{(\zeta^2 - 1)})\omega$. (2.10)

The response evidently depends on whether c is positive or negative and/or whether ζ is greater than, equal to, or less than unity. Usually c is positive, so we need consider only the other possibilities.

Case 1. $\zeta < 1$, that is, damping less than critical.
From equation (2.10),

$$s_{1,2} = -\zeta\omega \pm j\sqrt{(1-\zeta^2)}\,\omega \text{ where } j = \sqrt{-1},$$

so $$x = e^{-\zeta\omega t}\,[X_1 \exp(j\sqrt{(1-\zeta^2)}\omega t) + X_2 . \exp(j\sqrt{(1-\zeta^2)}\omega t)]$$

$$= Xe^{-\zeta\omega t} \sin(\sqrt{(1-\zeta^2)}\,\omega t + \phi) \ .$$

That is, the motion is an exponentially decaying harmonic as shown in Fig. 2.15.

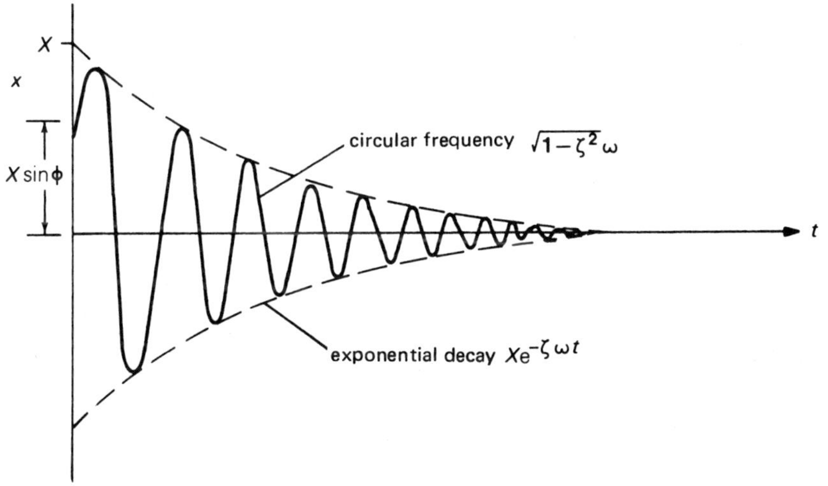

Fig. 2.15

The frequency of the viscously damped oscillation, ω_v is given by $\omega_v = \omega\sqrt{(1-\zeta^2)}$, that is, the frequency of oscillation is reduced by the damping action. However, in many systems this reduction is likely to be small, because very small values of ζ are common, for example in most engineering structures ζ is rarely greater than 0.02. Even if $\zeta = 0.2$, $\omega_v = 0.98\,\omega$. This is not true in those cases where ζ is large, for example in motor vehicles where ζ is typically 0.7 for shock absorbers.

Case 2. $\zeta = 1$, that is, critical damping
Both values of s are $-\zeta\omega$. However, two constants are required in the solution of equation (2.9), thus $x = (A + Bt)e^{-\omega t}$.

Critical damping represents the limit of periodic motion, hence the displaced body is restored to equilibrium in the shortest possible time, and without oscillation or overshoot. Many devices, particularly electrical instruments, are critically damped to take advantage of this property.

Case 3. $\zeta > 1$, that is, damping greater than critical
There are two real values of s, so $x = X_1 e^{s_1 t} + X_2 e^{s_2 t}$.

Since both values of s are negative the motion is the sum of two exponential decays as shown in Fig. 2.16.

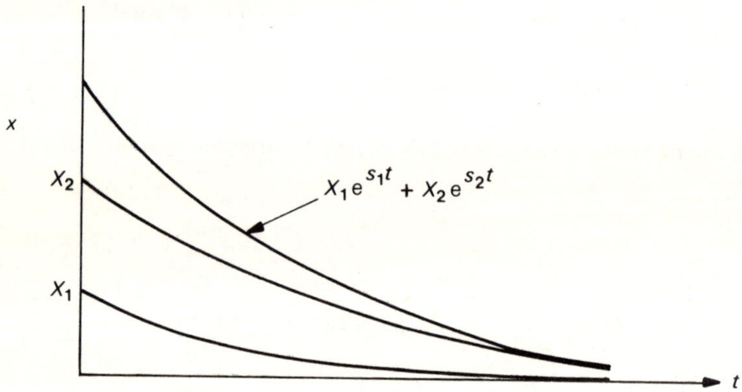

Fig. 2.16

Logarithmic Decrement Λ
A convenient way of determing the damping in a system is to measure the rate of decay of oscillation. It is usually not satisfactory to measure ω_v and ω, because unless $\zeta > 0.2$, $\omega \simeq \omega_v$.

Λ is the natural logarithm of the ratio of any two successive amplitudes in the same direction.

From Fig. 2.17, $\Lambda = ln(X_1/X_2)$

since $x = Xe^{-\zeta\omega t} \sin(\surd(1 - \zeta^2)\,\omega t + \phi)$

$$X_1 = Xe^{-\zeta\omega t} \text{ and } X_2 = Xe^{-\zeta\omega\,(t+\tau_v)}$$

that is, $\Lambda = ln\ Xe^{-\zeta\omega t}/Xe^{-\zeta\omega(t+\tau_v)} = \zeta\omega\,\tau_v$.

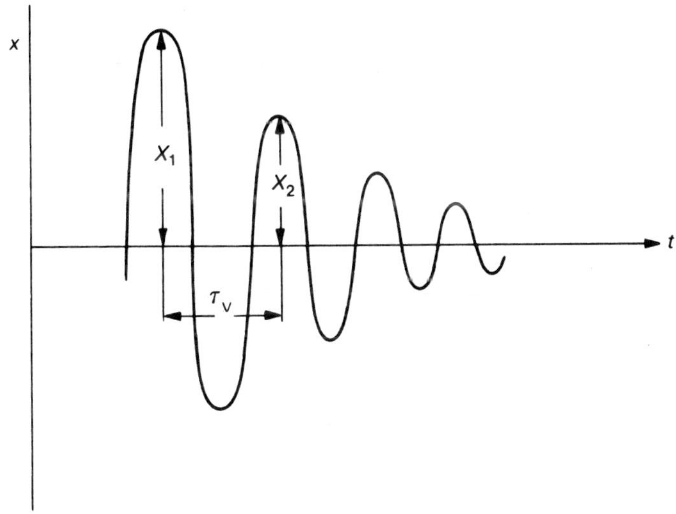

Fig. 2.17

Since $\tau_v = 2\pi/\omega_v = 2\pi\zeta/(\omega\sqrt{(1-\zeta^2)})$

$$\Lambda = 2\pi\zeta/\sqrt{(1-\zeta^2)}$$

For small values of $\zeta(\not>0.25)$, $\Lambda \cong 2\pi\zeta$.

Root Locus Study of Damping

It is often convenient to consider how the roots of equation (2.10) vary as ζ increases from zero. The roots of this equation are given by

$$s_{1,2}/\omega = -\zeta \pm j\sqrt{(1-\zeta^2)} \text{ for } 1 > \zeta > 0$$

and $s_{1,2}/\omega = -\zeta \pm \sqrt{(\zeta^2 - 1)}$ for $\zeta > 1$.

These roots can be conveniently displayed in a plot of Imaginary (s/ω) against Real (s/ω); since for every value of ζ there are two values of (s/ω), the roots when plotted form two loci as shown in Fig. 2.18. The position of a root in the (s/ω) plane indicates the frequency of oscillation (Im (s/ω) axis) of the system, if any, and the rate of growth or decay of oscillation (Re (s/ω) axis).

Because the Re (s/ω) is negative on the left of the Im (s/ω) axis, all roots which lie to the left of the Im (s/ω) axis represent a decaying oscillation and therefore a stable system. Roots to the right of the Im (s/ω) axis represent a growing oscillation and an unstable system.

This root locus study of damping is a useful design technique, because the effects of changing the damping ratio on the response of a system can easily be seen.

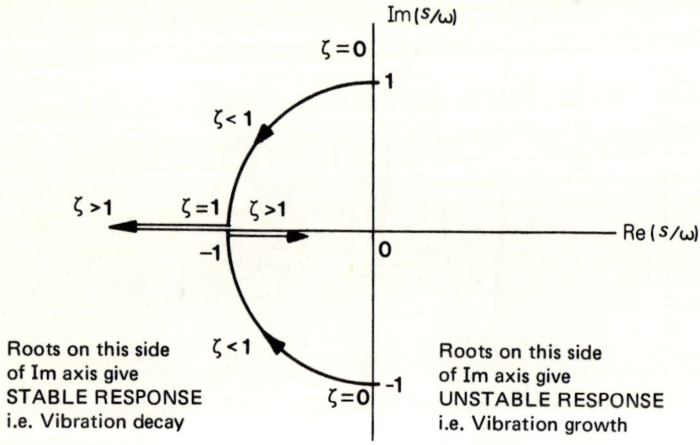

Fig. 2.18

2.2.2 Vibration with coulomb (dry friction) damping

Steady friction forces occur in many systems when relative motion takes place between adjacent members. These forces are independent of amplitude and frequency; they always oppose the motion and their magnitude may, to a first approximation, be considered constant. **Dry friction** can, of course, just be one of the damping mechanisms present; however, in some systems it is the main source of damping. In these cases the damping can be modelled as in Fig. 2.19.

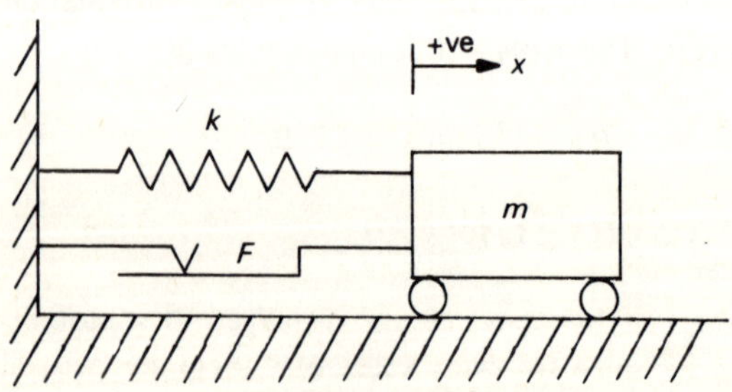

Fig. 2.19

The friction force F always opposes the motion, so that if the body is displaced a distance x_0 to the right and released from rest we have, for motion right to left only, $m\ddot{x} = F - kx$, or

$$m\ddot{x} + kx = F \quad . \tag{2.11}$$

The solution to the complementary function is $x = A \sin \omega t + B \cos \omega t$, and the complete solution is

$$x = A \sin \omega t + B \cos \omega t + (F/k) \quad . \tag{2.12}$$

Note. The particular integral may be found by using the D-operator. Thus equation (2.11) is

$$(\mathrm{D}^2 + \omega^2)x = F/m$$

so

$$x = (1/\omega^2)\,[1 + (\mathrm{D}^2/\omega^2)]^{-1}\,F/m$$

$$= [1 - (\mathrm{D}^2/\omega^2)\ldots.]\;F/m\omega^2 = F/k.$$

The initial conditions were $x = x_0$ at $t = 0$ and $\dot{x} = 0$ at $t = 0$. Substitution into equation (2.12) gives

$$A = 0 \text{ and } B = x_0 - (F/k),$$

that is,

$$x = (x_0 - (F/k)) \cos \omega t + (F/k). \tag{2.13}$$

At the end of the half cycle right to left, $\omega t = \pi$ and $x_{(t=\pi/\omega)} = -x_0 + (2F/k)$, i.e. a reduction in amplitude of $(2F/k)$ per half cycle.

From symmetry, for motion from left to right when the friction force acts in the opposite direction to the above, the initial displacement is $(x_0 - (2F/k))$ and the final displacement is therefore $(x_0 - (4F/k))$, i.e. the reduction in amplitude is $(4F/k)$ per cycle. This oscillation continues until the amplitude of the motion is so small that the spring force is unable to overcome the friction force F. This can happen whenever the amplitude is $\leqslant \pm (F/k)$. The manner of oscillation decay is shown in Fig. 2.20; the motion is sinusoidal for each half cycle, with successive half cycles centred on points distant $+ (F/k)$ and $- (F/k)$ from the origin. The oscillation ceases with $|x| \leqslant F/k$. The zone $x = \pm F/k$ is called the **Dead Zone.**

To determine the frequency of oscillation we rewrite the equation of motion (2.11) as

$$m\ddot{x} + k(x - (F/k)) = 0$$

Now if $x' = x - (F/k), \ddot{x}' = \ddot{x}$ so that $m\ddot{x}' + kx' = 0$ from which the frequency of oscillation is $(1/2\pi)\,\sqrt{(k/m)}$ Hz, that is, the frequency of oscillation is not affected by coulomb friction.

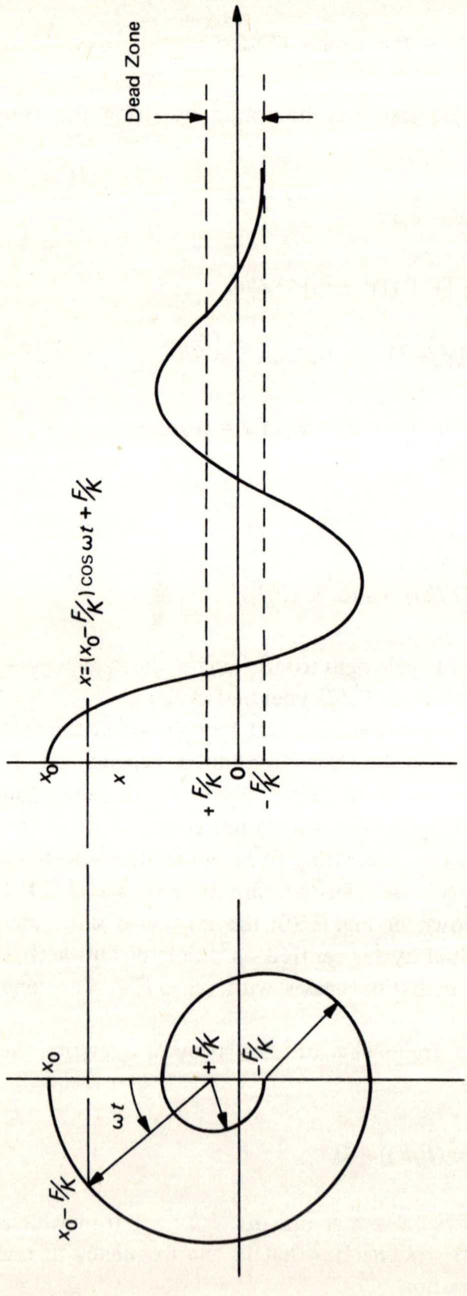

Fig. 2.20

2.2.3 Vibration with hysteretic damping

Experiements on the damping which occurs in solid materials when cyclically stressed have shown the damping to be independent of frequency. Fig. 2.21 shows a single degree of freedom model with hysteretic damping of coefficient c_H.

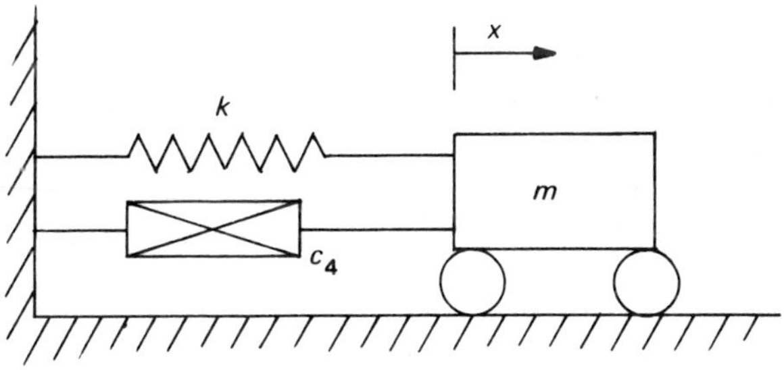

Fig. 2.21

The equation of motion is

$$m\ddot{x} + (c_H/\omega)\,\dot{x} + k\omega = 0$$

or $$m\ddot{x} + (k + jc_H)x = 0, \qquad \text{if } x = Xe^{j\omega t},$$

that is, the combined effect of the elastic and hysteretic resistance to motion can be represented as a complex stiffness.

Example 5

A single degree of freedom system has viscous damping, with $\zeta = 0.02$. Find the energy dissipated per cycle as a function of the energy in the system at the start of that cycle. Also find the amplitude of the 12th cycle if the amplitude of the 3rd cycle is 1.8 mm.

$$\zeta << 1, \text{ so } ln\ (X_1/X_2) = 2\pi\zeta = 0.126$$

$$X_1/X_2 = e^{0.126} = 1.134\ .$$

Energy at start of cycle $= \frac{1}{2}\,kX_1{}^2$
Energy at end of cycle $= \frac{1}{2}\,kX_2{}^2$

$$\frac{\text{Energy dissipated cycle}}{\text{Energy at start of cycle}} = \frac{\frac{1}{2}\,kX_1{}^2 - \frac{1}{2}\,kX_2{}^2}{\frac{1}{2}\,kX_1{}^2} = 1 - (X_2/X_1)^2$$

$$= 1 - 0.7773 = 0.223\ .$$

Now $$X_1/X_2 = 1.134, X_2/X_3 = 1.134 \ldots (X_{n-1})/X_n = 1.134.$$

Therefore $X_3/X_{12} = (1.134)^9 = 3.107$

i.e. $$X_{12} = \frac{1.8}{3.107} = 0.579 \text{ mm} .$$

Example 6

A gun is designed so that when fired the barrel recoils against a spring. At the end of the recoil a viscous damper is engaged which allows the barrel to return to its equilibrium position in the minimum time without overshoot. Determine the spring constant and damping coefficient for a gun whose barrel has a mass of 500 kg if the initial recoil velocity is 30 m/s and the recoil distance is 1.6 m.

Find also the time required for the barrel to return to a position 0.1 m from its initial position.

Since the initial kinetic energy of the barrel must equal the maximum strain energy in the spring,

$$\tfrac{1}{2}\, k\, 1.6^2 = \tfrac{1}{2}\, 500(30)^2$$

i.e. $k = 175$ kN/m.

Since critical damping is required, the damping coefficient is

$$2\sqrt{(500\ 175\ 10^3)} = 18\ 700 \text{ Ns/m}.$$

For critical damping $x = (A + Bt)e^{-\omega t}$

If $x = x_0$ and $\dot{x} = 0$ at $t = 0$

$$x = x_0(1 + \omega t)e^{-\omega t}.$$

Now $\omega = \sqrt{(175\ 10^3/500)} = 18.7$ rad/s and $x = 0.1$ m, so we require the solution to $0.1 = 1.6(1 + 18.7t)e^{-18.7t}$.

This equation can be solved in many ways, e.g. by trial and error or graphically, to give $t = 0.24$ s.

2.3 FORCED VIBRATION

Many real systems are subjected to periodic excitation. This may be due to unbalanced rotating parts, reciprocating components, or a shaking foundation. Sometimes large motions of the suspended body are desired as in vibratory feeders and compactors, but usually we require very low vibration amplitudes over a large range of exciting forces and frequencies. Some periodic forces are harmonic, but even if they are not, they can be represented as a series of harmonic functions using Fourier analysis techniques. Because of this the response of elastically supported bodies to harmonic exciting forces and motions must be studied.

2.3.1 Response of a viscous damped system to a simple harmonic exciting force with constant amplitude

Consider the system shown in Fig. 2.22. The body of mass m is connected by a spring and viscous damper to a fixed support whilst an harmonic force of frequency υ and amplitude F acts up on it, in the line of motion.

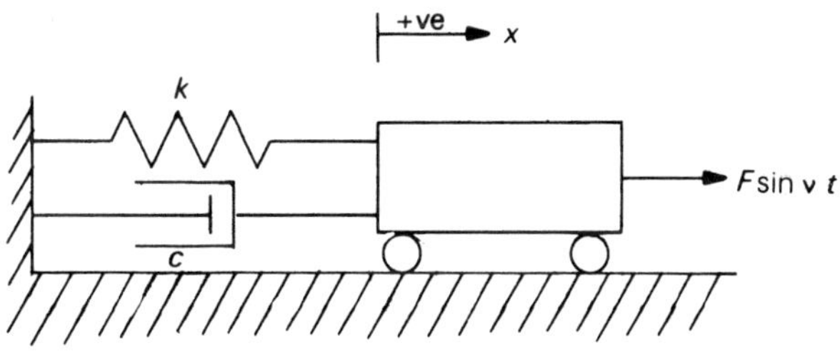

Fig. 2.22

The equation of motion is

$$m\ddot{x} + c\dot{x} + kx = F \sin \upsilon t \ . \tag{2.14}$$

The solution to $m\ddot{x} + c\dot{x} + kx = 0$ which has already been studied, is the complementary function; it represents the initial vibration which quickly dies away. The sustained motion is given by the particular solution. A solution $x = X(\sin\ \upsilon t - \phi)$ can be assumed, because this represents simple harmonic motion at the frequency of the exciting force with a displacement vector which lags the force vector by ϕ, i.e. the motion occurs after the application of the force.

Assuming $x = X \sin(\upsilon t - \phi)$,

$$\dot{x} = X\upsilon \cos(\upsilon t - \phi) = X\upsilon \sin(\upsilon t - \phi + \tfrac{1}{2}\pi))$$

and $$\ddot{x} = - X\upsilon^2 \sin(\upsilon t - \phi) = X\upsilon^2 \sin(\upsilon t - \phi + \pi)$$

The equation of motion (2.14) thus becomes

$$mX\upsilon^2 \sin(\upsilon t - \phi + \pi) + cX\upsilon \sin(\upsilon t - \phi + \pi/2) + kX \sin(\upsilon t - \phi) = F \sin\upsilon t \ .$$

A vector diagram of these forces can now be drawn, Fig. 2.23

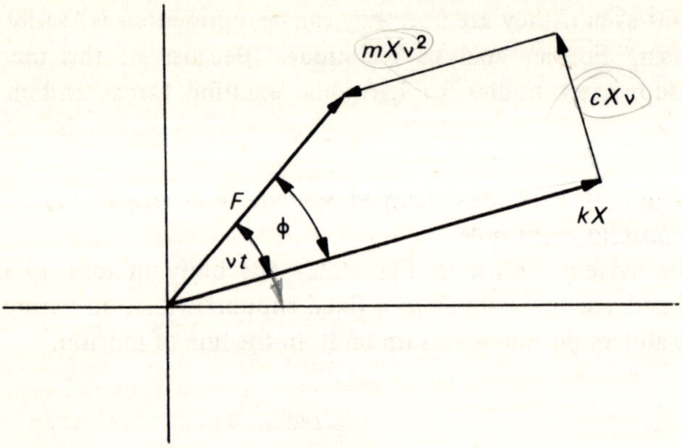

Fig. 2.23

From the diagram,

$$F^2 = (kX - mXv^2)^2 + (cXv)^2$$

or $$X = F/\sqrt{((k - mv^2)^2 + (cv)^2)} \tag{2.15}$$

and $$\tan\phi = cXv/(kX - mXv^2) \quad .$$

Thus the steady state solution to equation (2.14) is

$$x = F/\sqrt{((k - mv^2)^2 + (cv)^2)}\ \sin(vt - \tan^{-1} cv/(k - mv^2)).$$

The complete solution includes the transient motion given by the complementary function:

$$x = A\,e^{-\zeta\omega t}\sin(\omega\sqrt{(1 - \zeta^2}\,t) + \alpha) \quad .$$

Fig. 2.24 shows the combined motion.

Equation (2.15) can be written in a more convenient form if we put $\omega = \sqrt{(k/m)}$ and $X_s = F/k$. Then (2.15) becomes

$$X/X_s = 1/\sqrt{((1 - (v/\omega)^2)^2 + (2\zeta\, v/\omega)^2)} \tag{2.16}$$

and $$\tan\phi = 2\zeta(v/\omega)/(1 - (v/\omega)^2) \quad .$$

Since X_s is the static deflection of the system by a force F, X/X_s is known as the **dynamic magnification factor.**

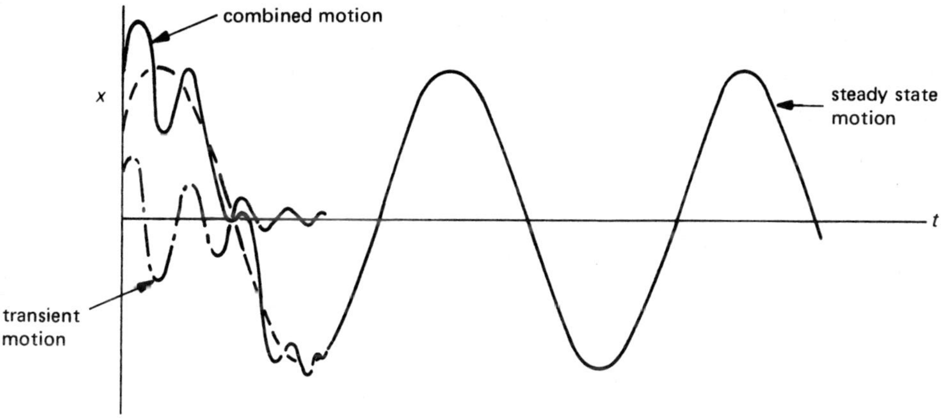

Fig. 2.24

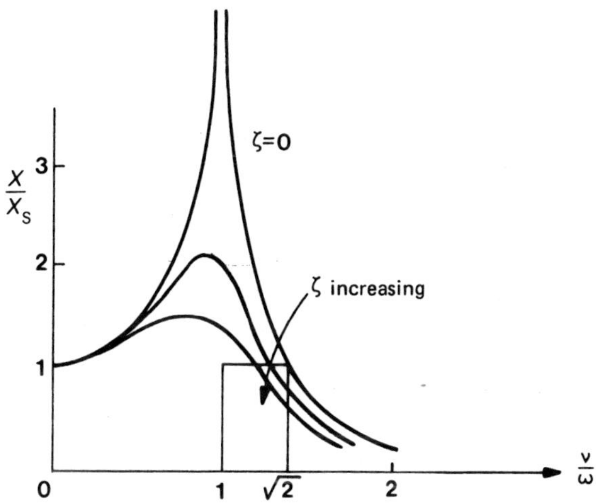

Fig. 2.25

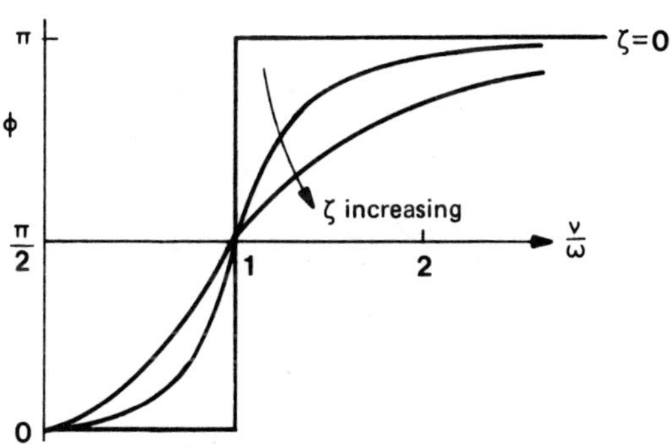

Fig. 2.26

By considering different values of the frequency ratio υ/ω, we can plot X/X_s and ϕ as functions of frequency for various values of ζ. Fig. 2.25 and 2.26 show the results.

The effect of the frequency ratio on the vector diagram is shown in Fig. 2.27.

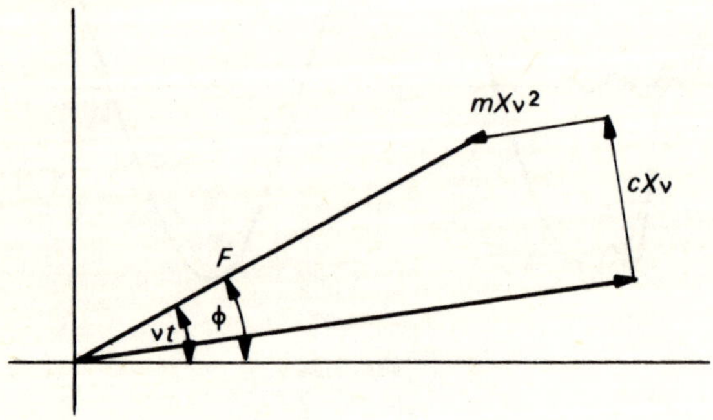

(a) $\upsilon/\omega << 1$ Exciting force approximately equal to spring force.

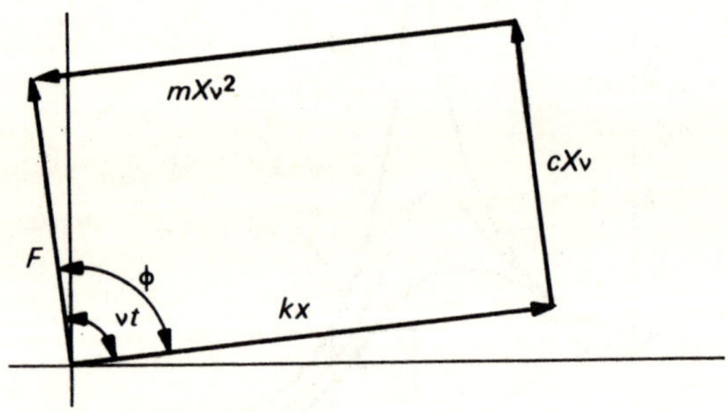

(b) $\upsilon/\omega = 1$ Exciting force equal to damping force, and inertia force equal to spring force.

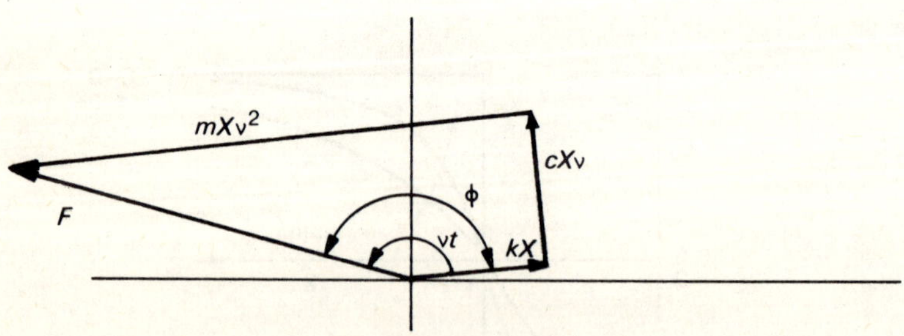

(c) $\upsilon/\omega >> 1$ Exciting force nearly equal to inertia force.

Fig. 2.27

The importance of mechanical vibrations arises mainly from the large values of X/X_s experienced in practice when υ/ω has a value near unity: this means that a small harmonic force can produce a large amplitude of vibration. The phenomenon known as **resonance** occurs when the forcing frequency is equal to the natural frequency, i.e. $\upsilon/\omega = 1$. The maximum value of X/X_s actually occurs at values of υ/ω less than unity: the value can be found by differentiating equation (2.16) with respect to υ/ω. Hence

$$(\upsilon/\omega)_{(X/X_s)\max} = \sqrt{(1 - 2\zeta^2)}$$

and $$(X/X_s)_{\max} = 1/(2\zeta\sqrt{(1 - \zeta^2)}).$$

For small values of ζ, $(X/X_s)_{\max} \cong 1/2\zeta$ which is the value pertaining to $\upsilon/\omega = 1$. $1/2\zeta$ is a measure of the damping in a system and is referred to as the Q factor.

Both *reciprocating and rotating unbalance* in a system produce an exciting force of the inertia type and result in the amplitude of the exciting force being proportional to the square of the frequency of excitation.

For an unbalanced body of mass m_r at an effective radius r, rotating at an angular speed υ, the exciting force is therefore $m_r . r . \upsilon^2$. If this force is applied to a single degree of freedom system such as that in Fig. 2.22, the component of the force in the direction of motion is $m_r r . \upsilon^2 . \sin\upsilon t$, and the amplitude of vibration is

$$X = \frac{(m_r/m) r\, (\upsilon/\omega)^2}{\sqrt{((1 - (\upsilon/\omega)^2)^2 + (2\zeta\, \upsilon/\omega)^2)}} \qquad (2.17)$$

(see equation (2.16)

The value of υ/ω for maximum X found by differentiating equation (2.17) is given by

$$(\upsilon/\omega)_{X\max} = 1/\sqrt{(1 - 2\zeta^2)}$$

i.e. the peak of the response curve occurs when $\upsilon > \omega$. This is shown in Fig. 2.28.

Also, $$X_{\max} = (m_r/m)\; r/2\zeta\sqrt{(1 - \zeta^2)}$$

It can be seen that away from the resonance condition ($\upsilon/\omega = 1$) the system response is not greatly affected by damping unless this happens to be large. Since in most mechanical systems the damping is small ($\zeta < 0.1$) it is often permissible to neglect the damping when evaluating the frequency for maximum amplitude and also the amplitude-frequency response away from the resonance condition.

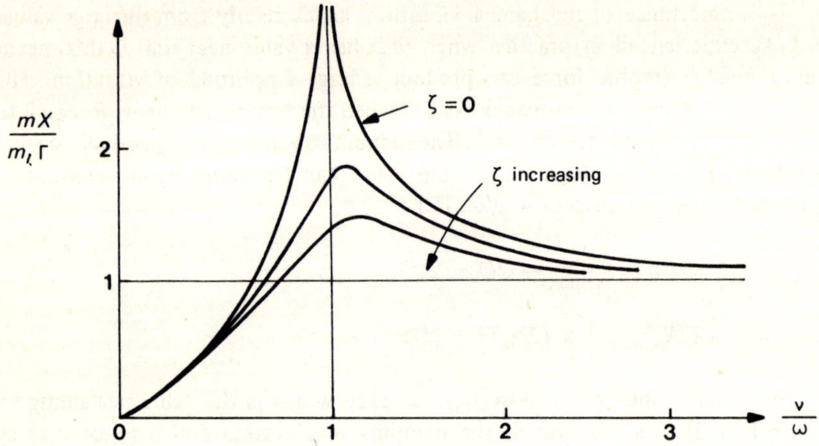

Fig. 2.28

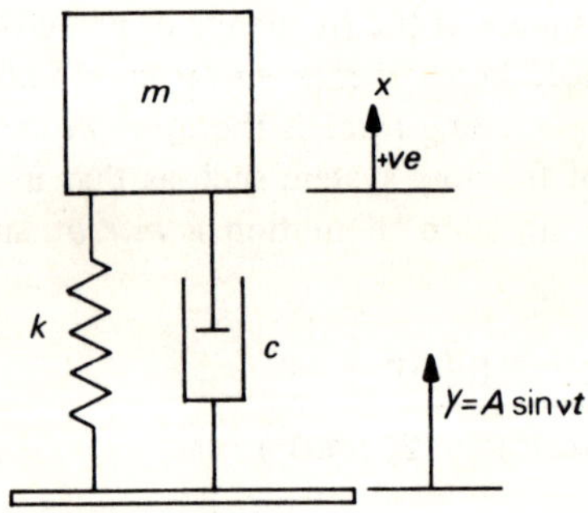

Fig. 2.29

2.3.2 Response of a viscous damped system supported on a foundation subjected to harmonic vibration

The system considered is shown in Fig. 2.29. The foundation is subjected to harmonic vibration A sinνt and it is required to determine the response, x, of the body. The equation of motion is

$$m\ddot{x} = c(\dot{y} - \dot{x}) + k(y - x) \tag{2.18}$$

If the displacement of the body relative to the foundation, u, is required, we may write $u = x - y$, and equation (2.18) becomes

$$m\ddot{u} + c\dot{u} + ku = -m\ddot{y} = m\nu^2 A \sin\nu t.$$

This equation is similar to (2.14) so that the solution may be written directly as

$$u = \frac{A(v/\omega)^2}{\sqrt{((1-(v/\omega)^2)^2 + (2\zeta\, v/\omega)^2)}} \sin\left\{ vt - \tan^{-1}\frac{2\zeta(v/\omega)}{1-(v/\omega)^2} \right\}$$

If the absolute motion of the body is required we rewrite equation (2.18) as

$$\begin{aligned} m\ddot{x} + c\dot{x} + kx &= c\dot{y} + ky \\ &= cAv\cos vt + kA\sin vt \\ &= \sqrt{((cv)^2 + k^2)}\, A\sin(vt + \alpha). \end{aligned}$$

Thus $$x = \frac{A\sqrt{(k^2 + (cv)^2)}}{\sqrt{((k-mv^2)^2 + (cv)^2)}} \sin\left(vt - \tan^{-1}\frac{cv}{k-mv^2} + \alpha\right).$$

The **motion transmissibility** is defined as the ratio of the amplitude of the body vibration to the amplitude of foundation vibration; that is:

$$X/A = \frac{\sqrt{(1 + (2\zeta(v/\omega)^2)}}{\sqrt{((1-(v/\omega)^2)^2 + (2\zeta\, v/\omega)^2)}}$$

Vibration isolation

The dynamic forces produced by machinery are often very large. The force transmitted to the foundation or supporting structure can be reduced by using flexible mountings with the correct properties; alternatively a machine can be isolated from foundation vibration by using the correct flexible mountings. Fig. 2.30 shows a model of such a system.

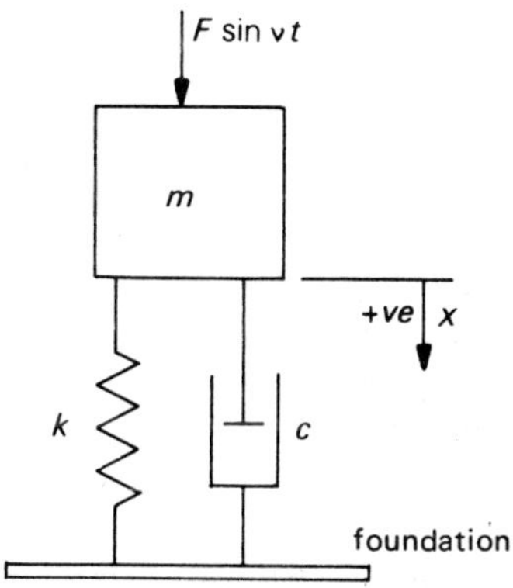

Fig. 2.30

The force transmitted to the foundation is the sum of the spring force and the damper force.

Transmitted force $= kx + c\dot{x}$, that is

Transmitted force amplitude, $F_T = \sqrt{((kX)^2 + (c\upsilon X)^2)}$.

The force transmission ratio or **transmissibility**, T_R, is given by

$$T_R = F_T/F = X\sqrt{(k^2 + (c\upsilon)^2)}/F$$

Since $X = (F/k)/\sqrt{((1 - (\upsilon/\omega)^2)^2 + (2\zeta\,\upsilon/\omega)^2)}$

$$T_R = \frac{\sqrt{(1 + (2\zeta\,\upsilon/\omega)^2)}}{\sqrt{((1 - (\upsilon/\omega)^2)^2 + (2\zeta\,\upsilon/\omega)^2)}},$$

that is, the force and motion transmissibilities are the same.

The effect of the frequency ratio on T_R is shown in Fig. 2.31.

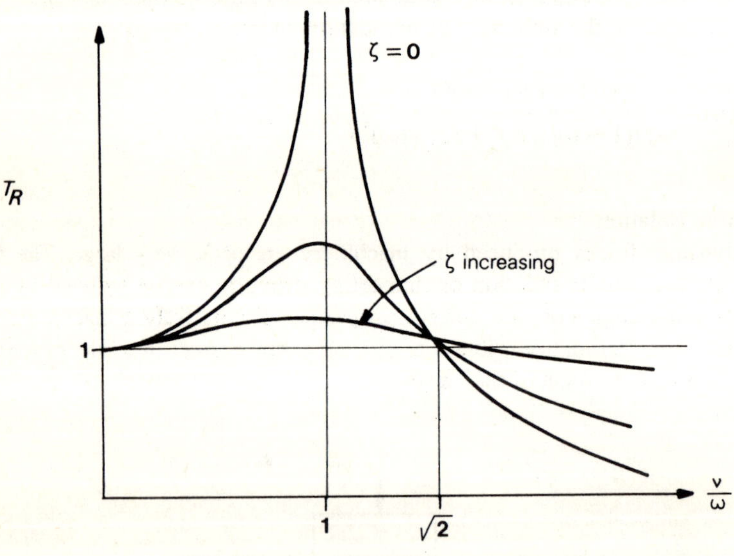

Fig. 2.31

It can be seen that for good isolation $\upsilon/\omega >> \sqrt{2}$, hence a low value of ω is required which implies a low stiffness, i.e. a flexible mounting: this may not always be acceptable in practice where a certain minimum stiffness is usually necessary to satisfy operating criteria.

Example 7

A single degree of freedom vibrational system of very small viscous damping ($\zeta < 0.1$) is excited by an harmonic force of frequency υ and amplitude F.

Show that the Q factor of the system is equal to the reciprocal of twice the damping ratio ζ.

It is sometimes difficult to measure ζ in this way because the static deflection of the body under a force F is very small. Another way is to obtain the two frequencies υ_1 and υ_2 (one either side of the resonance frequency ω) at the half-power points.

Show that $Q = 1/2\zeta = \omega/(\upsilon_2 - \upsilon_1)$

(The half-power points are those points on the response curve with an amplitude $1/\sqrt{2}$ times the amplitude at resonance).

$$X = (F/k)/\sqrt{((1-(\upsilon/\omega))^2 + (2\zeta\,\upsilon/\omega)^2)}$$

If $\upsilon = 0$ $X_{st} = F/k$, and

at resonance $\upsilon = \omega$, so $X = (F/k)/2\zeta = X_{st}/2\zeta$,

i.e. $Q = (X/X_{static})_{max} = 1/2\zeta$

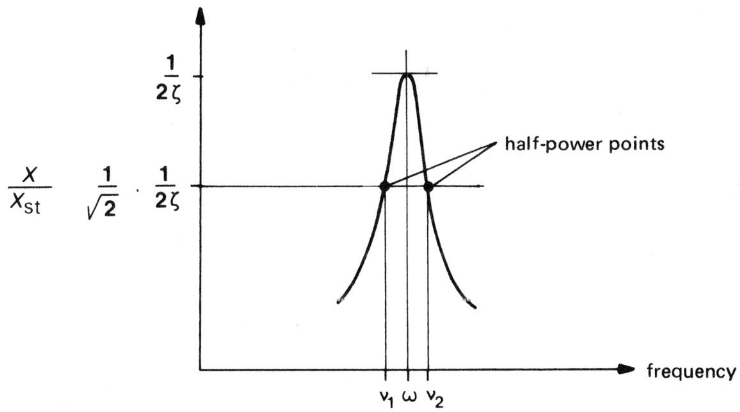

At half-power points,

$$(1/\sqrt{2}\ \ 1/2\zeta)^2 = \frac{1}{[1-(\upsilon/\omega)^2]^2 + [2\zeta\,\upsilon/\omega]^2} = \frac{1}{8\zeta^2}$$

Hence $(\upsilon/\omega)^4 + (\upsilon/\omega)^2\,(4\zeta^2 - 2) + (1 - 8\zeta^2) = 0$

$$\text{and } (\upsilon/\omega)^2 = \frac{(2-4\zeta^2) \pm \sqrt{((4\zeta^2-2)^2 - 4(1-8\zeta^2))}}{2}$$

If ζ is small, $\zeta^2 \cong \zeta^4 \cong 0$, i.e. $(\upsilon/\omega)^2 \cong 1 \pm 2\zeta$.

If roots are υ_1/ω and υ_2/ω,

$$(\upsilon_2{}^2 - \upsilon_1{}^2)/\omega^2 = 4\zeta = 2(\upsilon_2 - \upsilon_1)/\omega.$$

Thus $\quad 1/2\zeta = \omega/(\upsilon_2 - \upsilon_1) = Q.$

Example 8

A machine of mass 550 kg is flexibility supported on rubber mountings which provide a force proportional to displacement of 210 kN/m, together with a viscous damping force. The machine gives an exciting force of the form $R\upsilon^2 \cos \upsilon t$ where R is a constant. At very high speeds of rotation, the measured amplitude of vibration is 0.25 mm, and the maximum amplitude recorded as the speed is slowly increased from zero is 2 mm. Find the value of R and the damping ratio.

$$X = R\upsilon^2/(\sqrt{(k - m\upsilon^2)^2 + c^2\upsilon^2})$$

If υ is large, $X \to R\upsilon^2/(\sqrt{(m^2\upsilon^4)}) = R/m$.

$$R = mX = (550\ .\ 0.25)/1000 = 0.1375 \text{ kg m}.$$

For maximum X, $\mathrm{d}X/\mathrm{d}\upsilon = 0$, hence $\upsilon^2 = 2k^2/(2mk - c^2)$

and $$X_{\max} = \frac{R/2m}{\zeta\sqrt{(1 - \zeta^2)}}$$

$$\zeta\sqrt{(1 - \zeta^2)} = R/(2mX_{\max}) = 0.1375/(2.550.2.10^{-3})$$

$$= 0.062$$

i.e. $\quad \zeta = 0.062.$

CHAPTER 3

The vibrations of systems having more than one degree of freedom

Many real systems can be represented by a single degree of freedom model. However, most actual systems have several bodies and several restraints and therefore several degrees of freedom. The number of degrees of freedom that a system possesses is equal to the number of independent coordinates necessary to describe the motion of the system. Since no body is completely rigid, and no spring is without mass, every real system has more than one degree of freedom, and sometimes it is not sufficiently realistic to approximate a system by a single degree of freedom model. Thus, it is necessary to study the vibration of systems with more than one degree of freedom.

Each flexibly connected body in a multi-degree of freedom system can move independently of the other bodies, and only under certain conditions will all bodies undergo an harmonic motion at the same frequency. Since all bodies move with the same frequency, they all attain their amplitudes at the same time, even if they do not all move in the same direction. When such motion occurs the frequency is called a **natural frequency** of the system, and the motion is a principal mode of vibration: the number of natural frequencies and principal modes that a system possesses is equal to the number of degrees of freedom of that system. The deployment of the system at its lowest or first natural frequency is called its first mode, at the next highest or second natural frequency it is called the second mode and so on.

A two degree of freedom system will be considered initially. This is because the addition of more degrees of freedom increases the labour of the solution procedure but does not introduce any new analytical principles.

Initially, we will obtain the equations of motion for a two degree of freedom model, and from these find the natural frequencies and corresponding mode shapes.

3.1 THE VIBRATION OF SYSTEMS WITH TWO DEGREES OF FREEDOM

3.1.1 Free vibration of an undamped system

Some examples of two degree of freedom models are shown in Figs. 3.1(a) – (f).

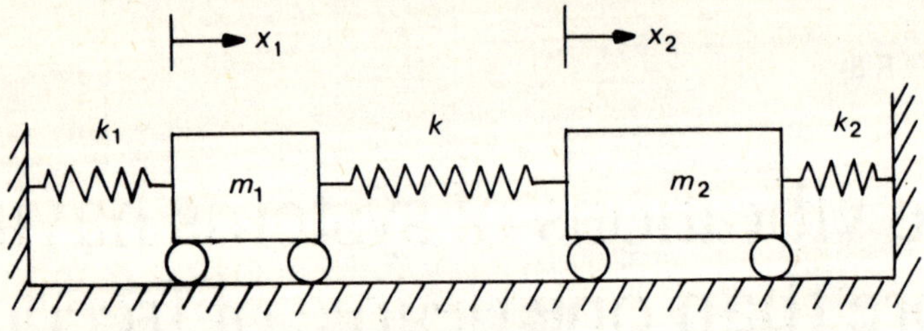

Fig. 3.1(a) – Linear undamped system, horizontal motion.

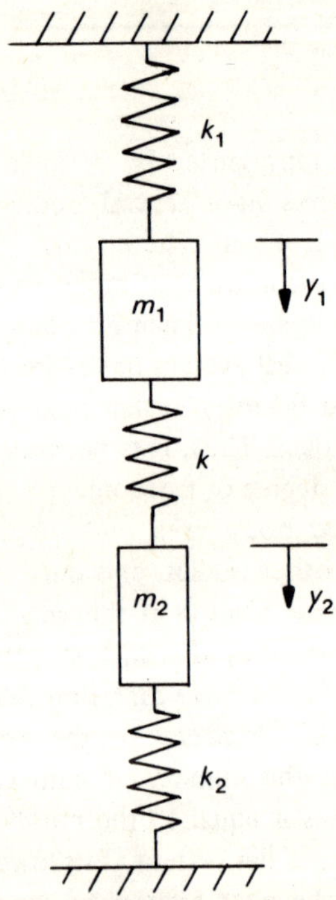

Fig. 3.1(b) – Linear undamped system, vertical motion.

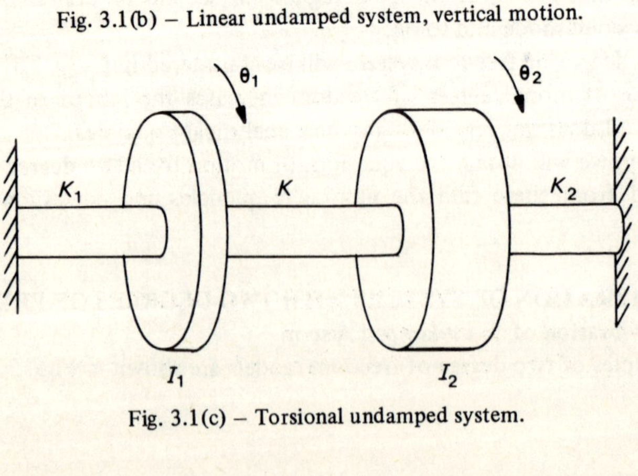

Fig. 3.1(c) – Torsional undamped system.

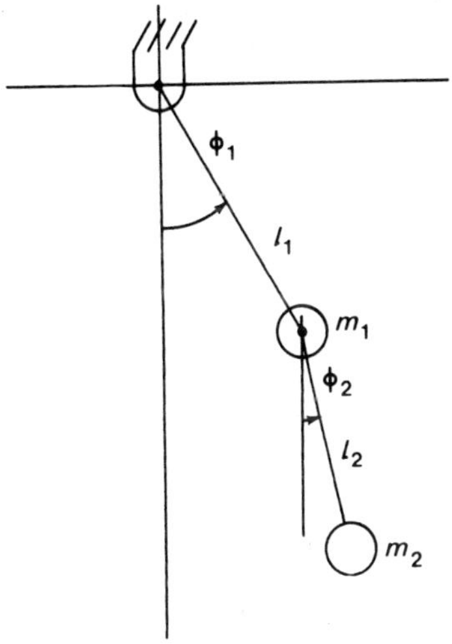

Fig. 3.1(d) – Coupled pendula.

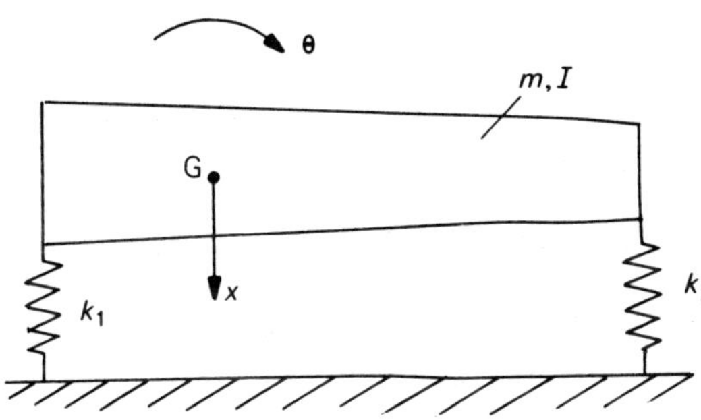

Fig. 3.1(e) – System with combined translation and rotation.

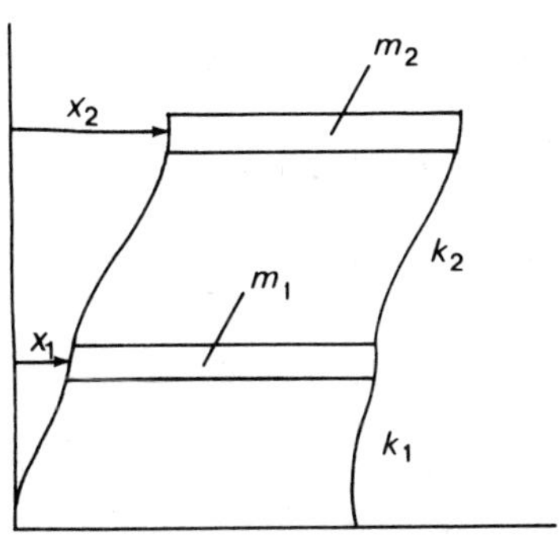

Fig. 3.1(f) – Shear frame.

Consider the system shown in Fig. 3.1(a). If $x_1 > x_2$ the FBDs are as shown in Fig. 3.2.

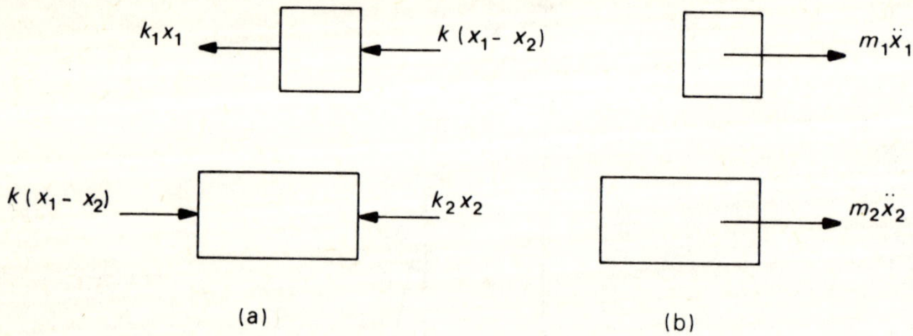

Fig. 3.2(a) – Applied forces. (b) Effective forces.

The equations of motion are

$$m_1\ddot{x}_1 = -k_1x_1 - k(x_1 - x_2) \tag{3.1}$$

$$m_2\ddot{x}_2 = k(x_1 - x_2) - k_2x_2 \tag{3.2}$$

The same equations are obtained if $x_1 < x_2$ is assumed because the direction of the central spring force is reversed.

Equations (3.1) and (3.2) can be solved for the natural frequencies and corresponding mode shapes by assuming a solution of the form $x_1 = A_1\sin(\omega t + \psi)$ and $x_2 = A_2\sin(\omega t + \psi)$.

This assumes that x_1 and x_2 oscillate with the same frequency ω and are either in phase or π out of phase. This is a sufficient condition to make ω a natural frequency.

Substituting these solutions into equations (3.1) and (3.2) gives

$$-m_1A_1\omega^2\sin(\omega t + \psi) = -k_1A_1\sin(\omega t + \psi) - k(A_1 - A_2)\sin(\omega t + \psi)$$

$$-m_2A_2\omega^2\sin(\omega t + \psi) = k(A_1 - A_2)\sin(\omega t + \psi) - k_2A_2\sin(\omega t + \psi)$$

i.e. $$A_1(k + k_1 - m_1\omega^2) + A_2(-k) = 0 \tag{3.3}$$

$$A_1(-k) + A_2(k_2 + k - m_2\omega^2) = 0\ . \tag{3.4}$$

A_1 and A_2 can be eliminated by writing

$$\begin{vmatrix} k + k_1 - m_1\omega^2 & -k \\ -k & k + k_2 - m_2\omega^2 \end{vmatrix} = 0\ . \tag{3.5}$$

This is the characteristic or frequency equation. Alternatively we may write

$$\left.\begin{aligned} A_1/A_2 &= k/(k + k_1 - m_1\omega^2) \quad \text{from (3.3)} \\ \text{and} \quad A_1/A_2 &= (k_2 + k - m_2\omega^2)/k \quad \text{from (3.4)} \end{aligned}\right\} \tag{3.6}$$

Thus $\quad k/(k + k_1 - m_1\omega^2) = (k_2 + k - m_2\omega^2)/k$

or

$$(k + k_1 - m_1\omega^2)\,(k_2 + k - m_2\omega^2) - k^2 = 0 \quad , \tag{3.7}$$

a result which could be obtained by multiplying out the determinant (3.5).

The solutions to equation (3.7) give the natural frequencies of free vibration for the system in Fig. 3.1(a). The corresponding mode shapes are found by substituting these frequencies, in turn, into either of equations (3.6).

Consider the case when $k_1 = k_2 = k$ and $m_1 = m_2 = m$. The frequency equation is $(2k - m\omega^2)^2 - k^2 = 0$; that is,

$$m^2\omega^4 - 4\,mk\omega^2 + 3k^2 = 0 \quad .$$

Thus $\omega_1 = \sqrt{(k/m)}$ rad/s and $\omega_2 = \sqrt{(3k/m)}$ rad/s .

If $\omega = \sqrt{(k/m)}$ rad/s, $(A_1/A_2)_{\omega = \sqrt{(k/m)}} = +1$

and if $\omega = \sqrt{(3k/m)}$ rad/s, $(A_1/A_2)_{\omega = \sqrt{(3k/m)}} = -1$

This gives the mode shapes corresponding to the frequencies ω_1 and ω_2. Thus, the first mode of free vibration occurs at a frequency $(1/2\pi)\sqrt{(k/m)}$ Hz and $(A_1/A_2)^{\mathrm{I}} = 1$, i.e. the bodies move in phase with each other and with the same amplitude as if connected by a rigid link, Fig. 3.3. The second mode of free vibration occurs at a frequency $(1/2\pi)\sqrt{(3k/m)}$ Hz and $(A_1/A_2)^{\mathrm{II}} = -1$, i.e. the bodies move exactly out of phase with each other, but with the same amplitude, see Fig. 3.3.

3.1.2 Free motion

The two modes of vibration can be written

$$\begin{Bmatrix} x_1 \\ x_2 \end{Bmatrix}^{\mathrm{I}} = \begin{Bmatrix} A_1 \\ A_2 \end{Bmatrix}^{\mathrm{I}} \sin(\omega_1 t + \psi_1)$$

and

$$\begin{Bmatrix} x_1 \\ x_2 \end{Bmatrix}^{\mathrm{II}} = \begin{Bmatrix} A_1 \\ A_2 \end{Bmatrix}^{\mathrm{II}} \sin(\omega_2 t + \psi_2)$$

where the ratio A_1/A_2 is specified for each mode.

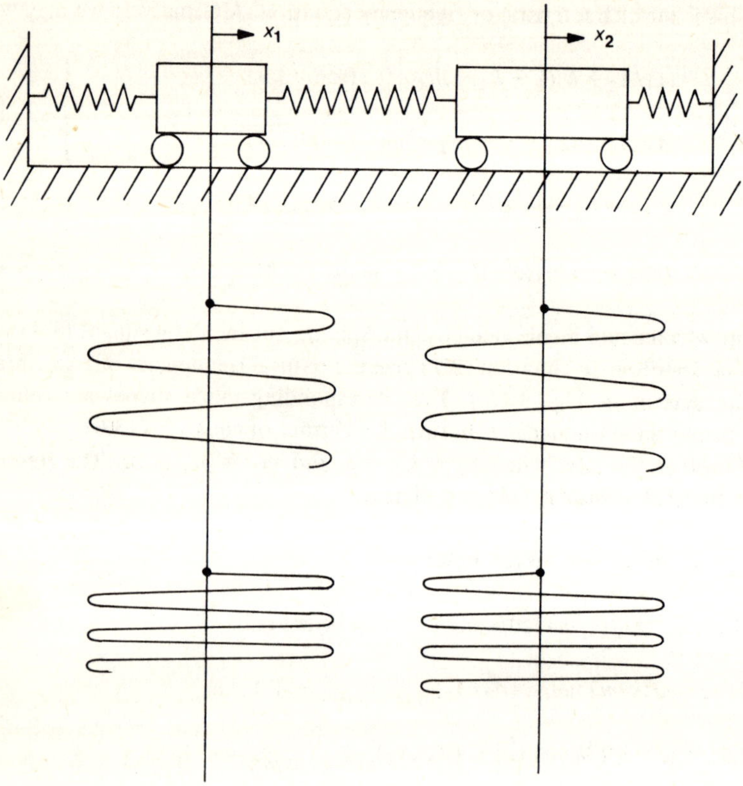

Fig. 3.3 – Natural frequencies and mode shapes.

Since each solution satisfies the equation of motion, the general solution is

$$\begin{Bmatrix} x_1 \\ x_2 \end{Bmatrix} = \begin{Bmatrix} A_1 \\ A_2 \end{Bmatrix}^{\mathrm{I}} \sin(\omega_1 t + \psi_1) + \begin{Bmatrix} A_1 \\ A_2 \end{Bmatrix}^{\mathrm{II}} \sin(\omega_2 t + \psi_2)$$

where A_1, A_2, ψ_1 and ψ_2 are found from the initial conditions.

For example, for the system considered above, if one body is displaced a distance X and released,

$$x_1(0) = X \text{ and } x_2(0) = \dot{x}_1(0) = \dot{x}_2(0) = 0$$

where $x_1(0)$ means the value of x_1 when $t = 0$, and similarly for $x_2(0)$, $\dot{x}_1(0)$ and $\dot{x}_2(0)$.

Now $x_1 = \sin(\sqrt{(k/m)}\,t + \psi_1) + \sin(\sqrt{(3k/m)}\,t + \psi_2)$

and $x_2 = \sin(\sqrt{(k/m)}\,t + \psi_1) - \sin(\sqrt{(3k/m)}\,t + \psi_2)$.

Substituting the initial conditions $x_1(0) = X$ and $x_2(0) = 0$ gives:

$$X = \sin\psi_1 + \sin\psi_2$$

and $0 = \sin\psi_1 - \sin\psi_2$

i.e. $\sin\psi_1 = \sin\psi_2 = \frac{1}{2}X$.

The remaining conditions give $\cos\psi_1 = \cos\psi_2 = 0$.

Hence $x_1 = (X/2)\cos\sqrt{(k/m)}\,t + (X/2)\cos\sqrt{(3k/m)}\,t$

and $x_2 = (X/2)\cos\sqrt{(k/m)}\,t - (X/2)\cos\sqrt{(3k/m)}\,t$,

i.e. both natural frequencies are excited and the motion of each body has two harmonic components.

3.1.3 Coordinate coupling

In some systems the motion is such that the coordinates are coupled in the equations of motion. Consider the system shown in Fig. 3.1(e), only motion in the plane of the figure is considered, horizontal motion being neglected because the lateral stiffness of the springs is assumed negligible. The coordinates of rotation, θ and translation, x, are coupled as shown in Fig. 3.4. G is the centre of mass of the rigid beam.

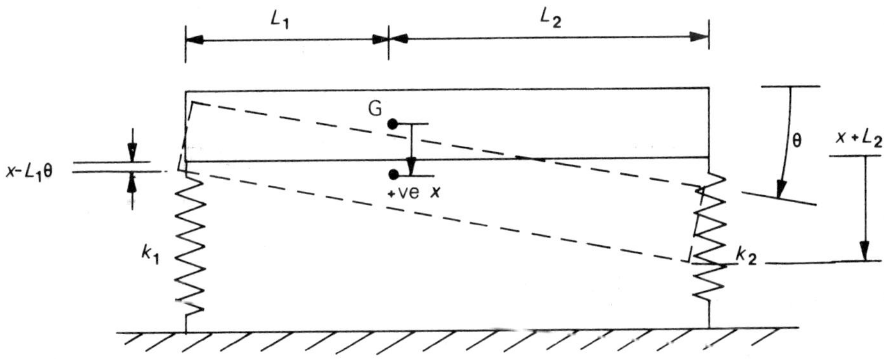

Fig. 3.4

The FBDs are shown in Fig. 3.5; since the weight of the beam is supported by the springs, both the initial spring forces and the beam weight may be omitted.

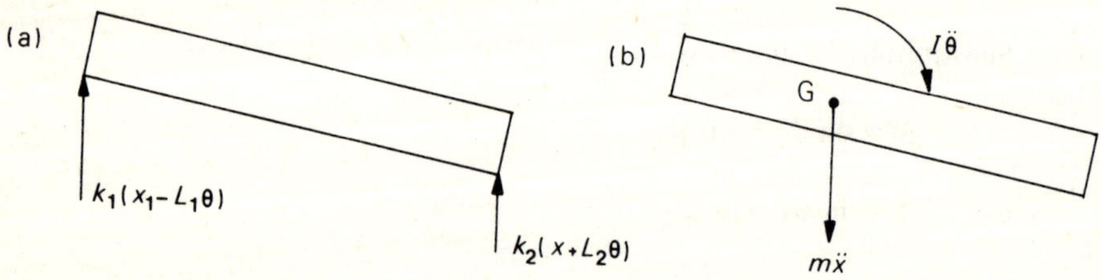

Fig. 3.5(a) – Applied forces. (b) Effective force and moment.

For small amplitudes of oscillation (so that $\sin\theta \simeq \theta$) the equations of motion are:

$$m\ddot{x} = -k_1(x - L_1\theta) - k_2(x + L_2\theta)$$

$$I\ddot{\theta} = k_1(x - L_1\theta)L_1 - k_2(x + L_2\theta)L_2 \ ,$$

i.e.

$$m\ddot{x} + (k_1 + k_2)x - (k_1L_1 - k_2L_2)\theta = 0$$

$$I\ddot{\theta} - (k_1L_1 - k_2L_2)x + (k_1L_1^2 + k_2L_2^2)\theta = 0 \ .$$

It will be noticed that these equations can be uncoupled by making $k_1L_1 = k_2L_2$; if this is arranged, translation (x – motion) and rotation (θ – motion) can take place independently. Otherwise translation and rotation occur simultaneously.

Assuming $x_1 = A_1 \sin(\omega t + \psi)$ and $\theta = A_2 \sin(\omega t + \psi)$, substituting into the equations of motion gives the frequency equation:

$$\begin{vmatrix} k_1 + k_2 - m\omega^2 & -(k_1L_1 - k_2L_2) \\ -(k_1L_1 - k_2L_2) & k_1L_1^2 + k_2L_2^2 - I\omega^2 \end{vmatrix} = 0$$

For each natural frequency, there is a corresponding mode shape.

3.1.4 Forced vibration

A two degree of freedom system which is excited by an harmonic force $F \sin\nu t$ is shown in Fig. 3.6. The equations of motion are:

$$m_1\ddot{x}_1 = -k_1x_1 - k(x_1 - x_2) + F \sin\nu t$$

$$m_2\ddot{x}_2 = k(x_1 - x_2) - k_2x_2 \ .$$

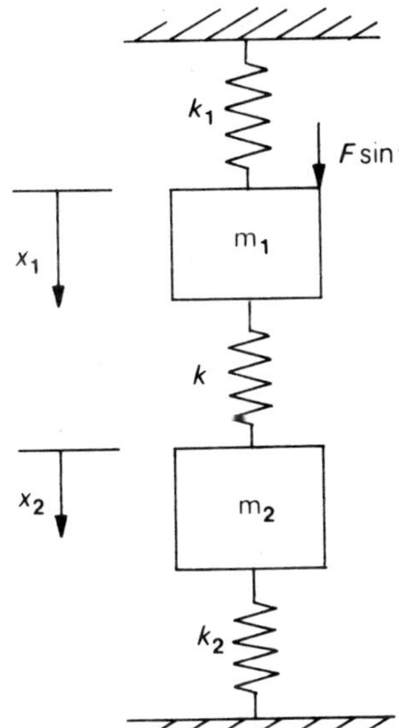

Fig. 3.6

Since there is zero damping, the motions are either in phase or π out of phase with the driving force so that a solution may be assumed:

$$x_1 = A_1 \sin \upsilon t \text{ and } x_2 = A_2 \sin \upsilon t \quad .$$

Substituting these solutions into the equations of motion gives:

$$A_1(k_1 + k - m_1\upsilon^2) + A_2(-k) = F$$

and $$A_2(-k) + A_2(k_2 + k - m_2\upsilon^2) = 0 \quad .$$

Thus $A_1 = F(k_2 + k - m_2\upsilon^2)/\Delta$

$$A_2 = Fk/\Delta$$

where $\Delta = (k_2 + k - m_2\upsilon^2)\,(k_1 + k - m_1\upsilon^2) - k^2$

and $\Delta = 0$ is the frequency equation.

Hence the system response is determined.

$A_1 = 0$ can be achieved by making $k_2 + k = m_2\upsilon^2$,

i.e. $\upsilon = \sqrt{((k_1 + k_2)/m_2)}$ rad/s. If $k_2 = 0$, $\upsilon = \sqrt{(k/m_2)}$ rad/s.

If the natural frequency of a single degree of freedom system is near to an exciting frequency, resonance is encountered. However, the addition of a further single degree of freedom system which is correctly tuned ($v = \sqrt{(k/m_2)}$), will reduce the amplitude of vibration to zero at this frequency. Such a device is known as a **dynamic vibration absorber.**

3.1.5 The undamped dynamic vibration absorber

If a single degree of freedom system or mode of a multi degree of freedom system is excited into resonance, large amplitudes of vibration result with accompanying high dynamic stresses and noise and fatigue problems. In most mechanical systems this is not acceptable.

If neither the excitation frequency nor the natural frequency can conveniently be altered, this resonance condition can often be successfully controlled by adding a further single degree of freedom system. Fig. 3.7 shows the primary or original system with an absorber added.

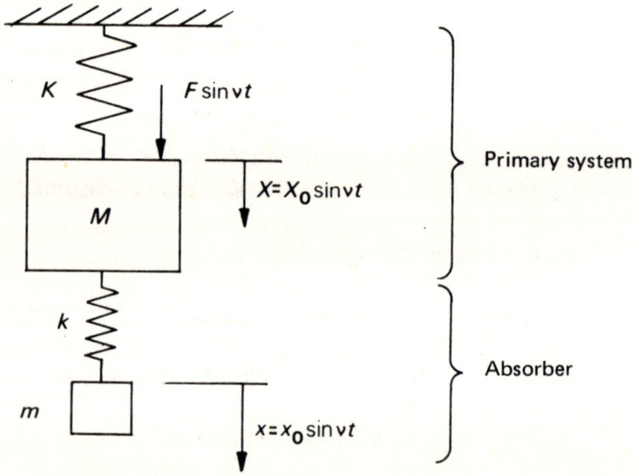

Fig. 3.7

From 3.1.4 above,

$$X_0 = F(k - mv^2)/\Delta$$

$$x_0 = Fk/\Delta$$

and $$\Delta = (K + k - Mv^2)(k - mv^2) - k^2 \ .$$

It can be seen that not only does the system now possess two natural frequencies Ω_1 and Ω_2 instead of one, but by arranging for $k - mv^2 = 0$, X_0 can be made zero.

Thus, if $\sqrt{(k/m)} = \sqrt{(K/M)}$, the response of the primary system at its original resonant frequency can be made zero. This is the usual tuning arrangment for an undamped absorber because the resonant problem in the primary system is only severe when $\upsilon \simeq \sqrt{(K/M)}$, see Fig. 3.8.

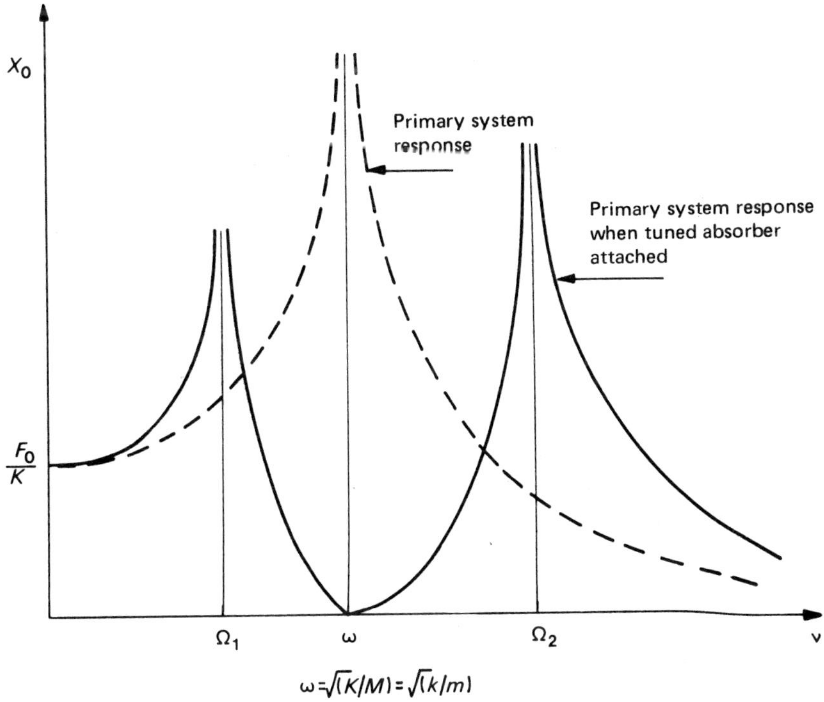

Fig. 3.8

When $X_0 = 0, x_0 = -F/k$, i.e. the force in the absorber spring (kx_0) is $-F$ so that the absorber applies a force to the primary system which is equal and opposite to the exciting force. Hence the body in the primary system has a net zero exciting force acting on it and therefore zero vibration amplitude.

If an absorber is correctly tuned, $\omega^2 = K/M = k/m$ and if the mass ratio $\mu = m/M$, the frequency equation $\Delta = 0$ is:

$$(\upsilon/\omega)^4 - (\upsilon/\omega)^2 (2 + \mu) + 1 = 0$$

$$\text{Hence } (\upsilon/\omega)^2 = (1 + (\mu/2)) \pm \sqrt{(\mu + (\mu^2/4))}$$

and the natural frequencies Ω_1 and Ω_2 are found.

$$\Omega_{1,2}/\omega = \{(1 + (\mu/2)) \pm \sqrt{(\mu + (\mu^2/4))}\}^{1/2} \quad .$$

For a small μ, Ω_1 and Ω_2 are very close to each other and near to ω; increasing μ gives better separation between Ω_1 and Ω_2, see Fig. 3.9.

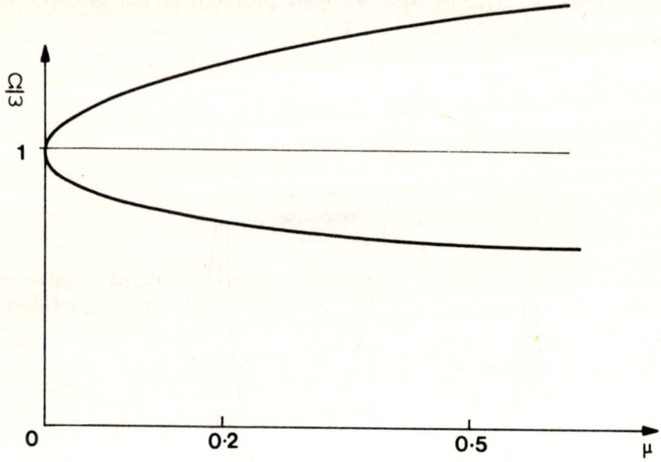

Fig. 3.9

This effect is of great importance in those systems where the excitation frequency may vary; if μ is small resonances at Ω_1 or Ω_2 may be excited.

Since $(\Omega_1/\omega)^2 = (1 + (\mu/2)) - \sqrt{(\mu + (\mu^2/4))}$

and $(\Omega_2/\omega)^2 = (1 + (\mu/2)) + \sqrt{(\mu + (\mu^2/4))}$

$$(\Omega_1{}^2\Omega_2{}^2)/\omega^4 = (1 + (\mu/2))^2 - (\mu + (\mu^2/4)) = 1\,,$$

i.e. $\Omega_1, \Omega_2 = \omega^2$.

Also $(\Omega_1/\omega)^2 + (\Omega_2/\omega)^2 = 2 + \mu$.

If the proximity of Ω_1 and Ω_2 to ω is likely to be a hazard, damping can be added in parallel with the absorber spring, to limit the response at these frequencies. Unfortunately, if damping is added, the response at frequency ω will no longer be zero. A design criterion that has to be carefully considered is the possible fatigue and failure of the absorber spring: this could have severe consequences. In view of this, some damped absorber systems dispense with the absorber spring and sacrifice some of the absorber effectiveness. This is particularly important in torsional systems, where the device is known as a **Lanchester damper.**

Example 9

A system is modelled as a straight link AB 1 m long of mass 10 kg, supported horizontally by springs A and B of stiffness k_1 and k_2 respectively. The moment of inertia of AB about its centre of mass G is 1 kg m^2 and G is located a distance a from A and b from B.

Find the relationship between k_1, k_2, a and b in order that one mode of free vibration shall be translational motion only and the other mode rotation only.

If $a = 0.3$ m and $k_1 = 13$ kN/m, find k_2 to give these two modes of vibration and calculate the two natural frequencies.

The frequency equation is (section 3.1.3)

$$\begin{vmatrix} k_1 + k_2 - m\omega^2 & -(k_1a - k_2b) \\ -(k_1a - k_2b) & k_1a^2 + k_2b^2 - I\omega^2 \end{vmatrix} = 0.$$

For the modes to be uncoupled $k_1a = k_2b$

and then $\omega_1 = \sqrt{(k_1 + k_2)/m}$ (bouncing mode)

and $\omega_2 = \sqrt{(k_1a^2 + k_2b^2)/I}$ (rocking mode) .

Since $a = 0.3$ m, $b = 0.7$ m and $k_1 = 13$ kN/m,

$$k_2 = (13.0.3)/0.7 = 5.57 \text{ kN/m} \quad .$$

$$\omega_1 = \sqrt{(18570/10)} = 43.1 \text{ rad/s}; f_1 = 6.86 \text{ Hz.}$$

$$\omega_2 = \sqrt{(13.0.3^2 + 5.57.0.7^2)10^3)} = 62.45 \text{ rad/s } f_2 = 9.94 \text{ Hz.}$$

Example 10

A two-wheel trailer is drawn over an undulating surface such that the vertical motion of the tyre may be regarded as sinusoidal, the pitch of the undulations being 5 m. The combined stiffness of the tyres is 170 kN/m and that of the main springs is 60 kN/m, the axle and attached parts have a mass of 400 kg, and the mass of the body is 500 kg. Find the critical speeds of the trailer in km/h.

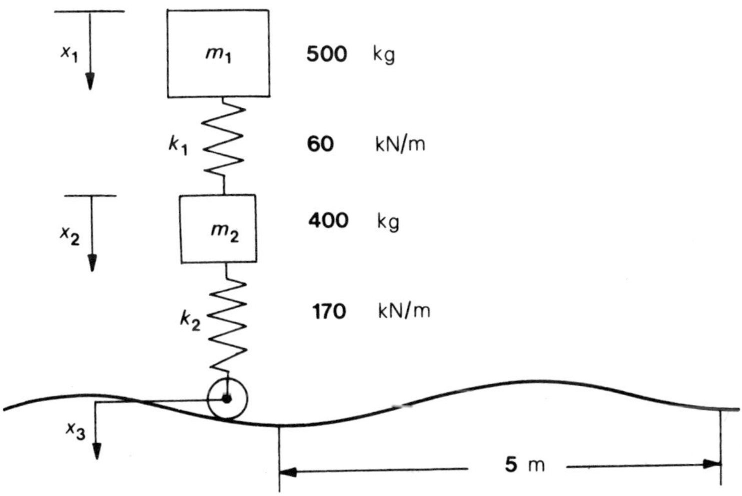

The equations of motion are:

$$m_1\ddot{x}_1 = -k_1(x_1 - x_2)$$
$$m_2\ddot{x}_2 = k_1(x_1 - x_2) - k_2(x_2 - x_3).$$

Assuming $x_1 = A_1 \sin \upsilon t$, $x_2 = A_2 \sin \upsilon t$, $x_3 = A_3 \sin \upsilon t$,

$$A_1(k_1 - m_1\upsilon^2) + A_2(-k_1) = 0$$

$$A_1(-k_1) + A_2(k_1 + k_2 - m_2\upsilon^2) = k_2A_3 \quad .$$

The frequency equation is:

$$(k_1 + k_2 - m_2\upsilon^2)(k_1 - m_1\upsilon^2) - k_1^2 = 0 \quad .$$

The critical speeds are those which correspond to the natural frequencies and hence excite resonances.

The frequency equation is:

$$m_1m_2\upsilon^4 - \upsilon^2(m_1k_1 + m_1k_2 + m_2k_1) + k_1k_2 = 0 \quad .$$

Hence

$$(500 \times 400)\,\upsilon^4 - \upsilon^2((500 \times 60) + (500 \times 170) + (400 \times 60))\,10^3 + (60 \times 170 \times 10^6) = 0 \quad .$$

i.e. $0.2\,\upsilon^4 - 139\,\upsilon^2 + 10\,200 = 0$

Thus $\upsilon = 16.3$ rad/s or 20.78 rad/s

and $f = 2.59$ Hz or 3.3 Hz.

Now if the trailer is drawn at V km/h,

or $V/3.6$ m/s; the frequency is $V/(3.6 \times 5)$ Hz.

Therefore the critical speeds are:

$$V_1 = 18 \times 2.59 = 46.6 \text{ km/h}$$

$$V_2 = 18 \times 3.3 = 59.4 \text{ km/h}.$$

Towing the trailer at either of these speeds will excite resonance in the system.

Example 11

A system has a violent resonance at 79 Hz. As a trial remedy a vibration absorber is attached which results in a resonance frequency of 65 Hz. How many such absorbers are required if no resonance is to occur between 60 and 120 Hz?

From section 3.1.5,

$$(\Omega_1/\omega)^2 + (\Omega_2/\omega)^2 = 2 + \mu, \text{ and } \Omega_1 . \Omega_2 = \omega^2 .$$

In the case of 1 absorber, $\omega = 79$ Hz, $\Omega_1 = 65$ Hz and $\Omega_2 = 79^2/65 = 96$ Hz.

Thus $(65/79)^2 + (96/79)^2 = 2 + \mu$ and $\mu = 0.154$.

In the case of n absorbers, if $\Omega_1 = 60$ Hz, $\Omega_2 = 79^2/60 = 104$ Hz(too low).

So require $\Omega_2 = 120$ Hz and $\Omega_1 = 79^2/120 = 52$ Hz

Hence $(52/79)^2 + (120/79)^2 = 2 + \mu'$.

$\mu' = 0.74 = \text{n}.\mu$

$\text{n} = 0.74/0.154 = 4.82$.

Thus 5 absorbers are required.

3.1.6 System with viscous damping

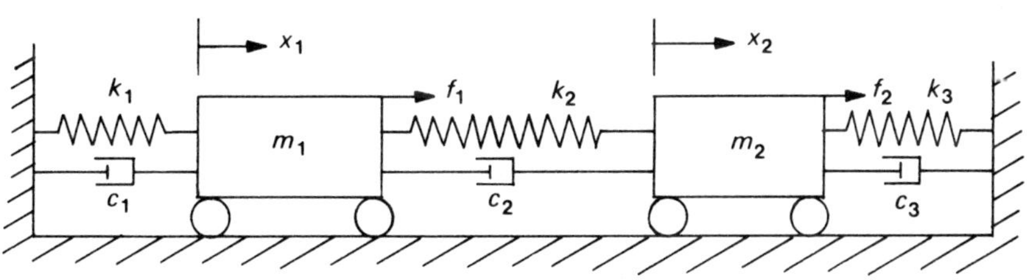

Fig. 3.10

For the system shown in Fig. 3.10 the equations of motion are:

$$m_1\ddot{x}_1 + k_1x_1 + k_2(x_1 - x_2) + c_1\dot{x}_1 + c_2(\dot{x}_1 - \dot{x}_2) = f_1$$

$$m_2\ddot{x}_2 + k_2(x_2 - x_1) + k_3x_2 + c_2(\dot{x}_2 - \dot{x}_1) + c_3\dot{x}_2 = f_2 \ .$$

The assumed solutions are $x_1 = A_1 e^{st}$ and $x_2 = A_2 e^{st}$ where $s = a + jb$, $j = \sqrt{-1}$ and a and b are real; i.e. each solution contains an harmonic and a decay component. Substituting these solutions into the equations of motion gives a frequency equation of the form $s^4 + \alpha s^3 + \beta s^2 + \gamma s + \delta = 0$.

From this equation four roots and four values for s are obtained. In general, because the coefficients α, β, γ and δ are real, the roots form two complex conjugate points such as $a_1 \pm jb_1$ and $a_2 \pm jb_2$. These represent solutions of the form $x = \text{Re}\, X\, e^{at} . e^{jbt} = Xe^{at} . \cos bt$.

It is often convenient to plot these roots on a complex plane as shown in Fig. 3.11. This is known as the s-plane.

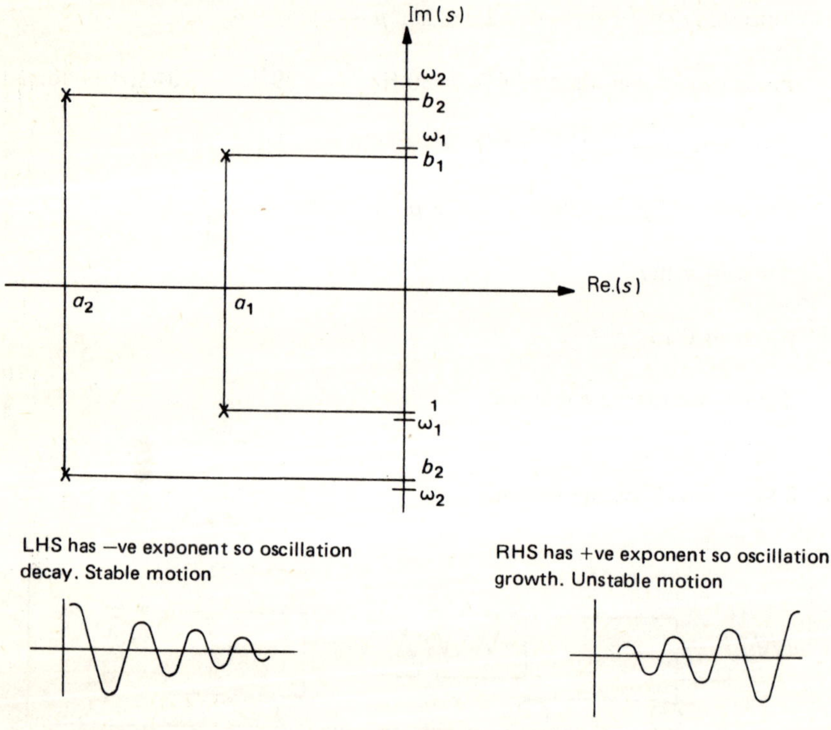

Fig. 3.11

For light damping the damped frequency for each mode is approximately equal to the undamped frequency, i.e. $b_1 \triangleq \omega_1$ and $b_2 \triangleq \omega_2$.

The right-hand side of the s-plane (Re(s) +ve) represents a root with a positive exponent, i.e. a term which grows with time, so unstable motion exists. The left-hand side contains roots with a negative exponent so stable motion exists. See also Fig. 2.18.

All passive systems have negative real parts and are therefore stable, but some systems such as rolling wheels and rockets can become unstable, and thus it is important that the stability of a system is considered. This can be conveniently done by plotting the roots of the frequency equation on the s-plane.

3.2 THE VIBRATION OF SYSTEMS WITH MORE THAN TWO DEGREES OF FREEDOM

The vibration analysis of systems with three or more degrees of freedom can be carried out in the same way as the analysis given above for two degrees of freedom. However, the method becomes tedious for many degrees of freedom, and numerical methods may have to be used to solve the frequency equation. A computer can of course be used to solve the frequency equation and determine the corresponding mode shapes. However, considerable economies in writing the analysis and performing the computations can be achieved, by adopting a matrix method for the analysis. For a full description of this method see *Mechanical Vibrations: An Introduction to Matrix Methods* by J. M. Prentis and F. A. Leckie. Longmans.

Example 12

Determine the highest natural frequency and associated mode shape for torsional vibrations of the system shown in the diagram.

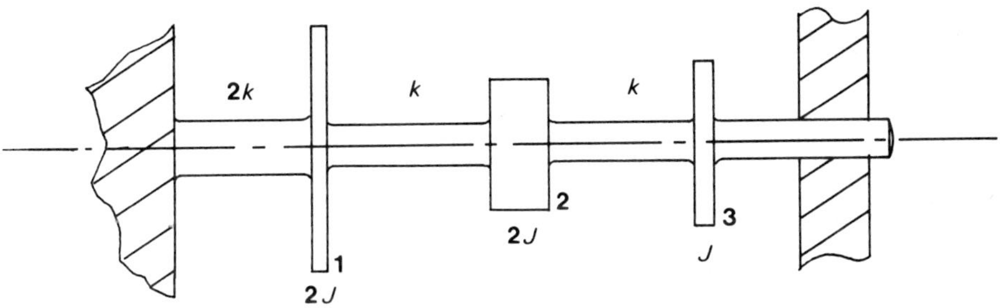

The shaft is assumed to be light. If the rotations of 1, 2, 3 are θ_1, θ_2, θ_3 respectively the equations of motion are:

$$2J\ddot{\theta}_1 + 2k\theta_1 + k(\theta_1 - \theta_2) = 0$$
$$2J\ddot{\theta}_2 + k(\theta_2 - \theta_1) + k(\theta_2 - \theta_3) = 0$$
$$J\ddot{\theta}_3 + k(\theta_3 - \theta_2) = 0 \quad .$$

If θ_1, θ_2 and θ_3 take the form $\theta\sin\omega t$ and $\lambda = (J\omega^2)/k$, these equations can be written

$$\begin{aligned} \tfrac{3}{2}\,\theta_1 - \tfrac{1}{2}\,\theta_2 \qquad\qquad &= \lambda\theta_1 \\ \tfrac{1}{2}\,\theta_1 + \theta_2 - \tfrac{1}{2}\,\theta_3 &= \lambda\theta_2 \\ -\theta_2 + \theta_3 &= \lambda\theta_3 \end{aligned}$$

i.e. $$\begin{bmatrix} 1.5 & -0.5 & 0 \\ 0.5 & 1 & -0.5 \\ 0 & -1 & 1 \end{bmatrix} \begin{Bmatrix} \theta_1 \\ \theta_2 \\ \theta_3 \end{Bmatrix} = \lambda \begin{Bmatrix} \theta_1 \\ \theta_2 \\ \theta_3 \end{Bmatrix}$$

or $[s]\ \{\theta\} = \lambda\ \{\theta\}$

where $[s]$ is the system matrix, $\{\theta\}$ is a column matrix, and the factor λ is a scalar quantity.

This equation can be solved by an iteration procedure. This procedure is started by assuming a set of deflections for the column matrix $\{\theta\}$ and multiplying by $[s]$: this results in a new column matrix. This matrix is normalised by making one of the amplitudes unity and dividing each term of the column by the particular amplitude which was put equal to unity. The procedure is repeated until the amplitudes stabilise, to a definite pattern. Convergence is always to the highest value of λ and its associated column matrix. Since $\lambda = (J\omega^2)/k$, this means that the highest natural frequency is found. Thus to start the iteration a reasonable assumed mode would be:

$$\begin{Bmatrix} \theta_1 \\ \theta_2 \\ \theta_3 \end{Bmatrix} = \begin{Bmatrix} 1 \\ -1 \\ 2 \end{Bmatrix}$$

Now, $$\begin{bmatrix} 1.5 & -0.5 & 0 \\ -0.5 & 1 & -0.5 \\ 0 & -1 & 1 \end{bmatrix} \begin{Bmatrix} 1 \\ -1 \\ 2 \end{Bmatrix} = \begin{Bmatrix} 2 \\ -2.5 \\ 3 \end{Bmatrix} = 3 \begin{Bmatrix} 0.67 \\ -0.83 \\ 1 \end{Bmatrix}.$$

Using this new column matrix gives;

$$\begin{bmatrix} 1.5 & -0.5 & 0 \\ -0.5 & 1 & -0.5 \\ 0 & 1 & 1 \end{bmatrix} \begin{Bmatrix} 0.67 \\ -0.83 \\ 1.00 \end{Bmatrix} = \begin{Bmatrix} 1.415 \\ -1.665 \\ 1.83 \end{Bmatrix} = 1.83 \begin{Bmatrix} 0.77 \\ -0.91 \\ 1.0 \end{Bmatrix}$$

and eventually, by repeating the process the following is obtained;

$$\begin{bmatrix} 1.5 & -0.5 & 0 \\ -0.5 & 1 & -0.5 \\ 0 & 1 & 1 \end{bmatrix} \begin{Bmatrix} 1 \\ -1 \\ 1 \end{Bmatrix} = 2 \begin{Bmatrix} 1 \\ -1 \\ 1 \end{Bmatrix}.$$

Hence $\lambda = 2$ and $\omega^2 = 2k/J$. λ is an **eigenvalue** of $[s]$, and the associated value of $\{\theta\}$ is the corresponding **eigenvector** of $[s]$. The eigenvector gives the mode shape.

The lowest eigenvalue can be found by inverting $[s]$ and premultiplying $[s]\ \{\theta\} = \lambda\ \{\theta\}$ by $\lambda^{-1}\ [s]^{-1}$.

Thus $[s]^{-1}\ \{\theta\} = \lambda^{-1}\ \{\theta\}$. Iteration of this equation yields the largest value of λ^{-1} and hence the lowest natural frequency. A reasonable assumed mode for the first iteration would be;

$$\begin{Bmatrix} 1 \\ 1 \\ 2 \end{Bmatrix}.$$

Alternatively, the lowest eigenvalue can be found directly from the flexibility matrix. The flexibility matrix is written in terms of the influence coefficients. The influence coefficient α_{pq} of a system is the deflection (or rotation) at the point p due to a unit force (or moment) applied at point q. Thus

$$\theta_1 = \alpha_{11}\, 2J\, \theta_1 \omega^2 + \alpha_{12}\, 2J\, \theta_2 \omega^2 + \alpha_{13}\, J\, \theta_3 \omega^2$$

$$\theta_2 = \alpha_{21}\, 2J\, \theta_1 \omega^2 + \alpha_{22}\, 2J\, \theta_2 \omega^2 + \alpha_{23}\, J\, \theta_3 \omega^2$$

$$\theta_3 = \alpha_{31}\, 2J\, \theta_1 \omega^2 + \alpha_{32}\, 2J\, \theta_2 \omega^2 + \alpha_{33}\, J\, \theta_3 \omega^2$$

or

$$\begin{bmatrix} 2\alpha_{11} & 2\alpha_{12} & \alpha_{13} \\ 2\alpha_{21} & 2\alpha_{22} & \alpha_{23} \\ 2\alpha_{31} & 2\alpha_{32} & \alpha_{33} \end{bmatrix} \begin{Bmatrix} \theta_1 \\ \theta_2 \\ \theta_3 \end{Bmatrix} = \frac{1}{J\omega^2} \begin{Bmatrix} \theta_1 \\ \theta_2 \\ \theta_3 \end{Bmatrix}.$$

The influence coefficients are calculated by applying a unit force or moment to each body in turn. Since the same unit force acts between the support and its point of application, the displacement of the point of application of the force is the sum of the extensions of the springs extended. The displacements of all points beyond the point of application of the force have the same displacement.

Thus $\alpha_{11} = \alpha_{12} = \alpha_{13} = \alpha_{21} = \alpha_{31} = \dfrac{1}{2k}$

$$\alpha_{22} = \alpha_{23} = \alpha_{32} = \frac{1}{2k} + \frac{1}{k} = \frac{3}{2k}$$

$$\alpha_{33} = \frac{1}{2k} + \frac{1}{k} + \frac{1}{k} = \frac{5}{2k}$$

Iteration converges to the highest value of $k/J\omega^2$ and hence to the lowest natural frequency.

Orthogonality of the Principal Modes of Vibration

Consider a linear elastic system that has n degrees of freedom, n natural frequencies, and n principal modes.

The orthogonality relation between the principal modes of vibration for an n degree of freedom system is;

$$\sum_{i=1}^{n} m_i\, A_i(r)\, A_i(s) = 0$$

where $A_i(r)$ are the amplitudes corresponding to the rth mode and $A_i(s)$ are the amplitudes corresponding to the sth mode.

This relationship is used to sweep unwanted modes from the system matrix, as illustrated in the following example.

Example 13
Consider the three degree of freedom system shown.

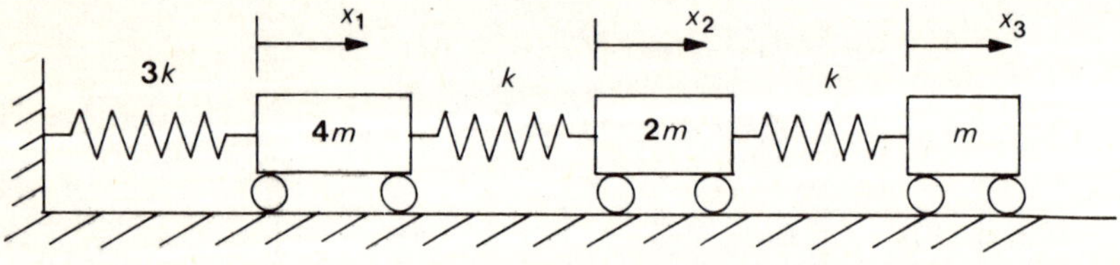

The equations of motion in terms of the influence coefficients are:

$$\begin{aligned} x_1 &= 4\alpha_{11}\, m x_1 \omega^2 + 2\alpha_{12}\, m x_2 \omega^2 + \alpha_{13}\, m x_3 \omega^2 \\ x_2 &= 4\alpha_{21}\, m x_1 \omega^2 + 2\alpha_{22}\, m x_2 \omega^2 + \alpha_{23}\, m x_3 \omega^2 \\ x_3 &= 4\alpha_{31}\, m x_1 \omega^2 + 2\alpha_{32}\, m x_2 \omega^2 + \alpha_{33}\, m x_3 \omega^2 \end{aligned}$$

i.e. $$\begin{Bmatrix} x_1 \\ x_2 \\ x_3 \end{Bmatrix} = \omega^2 m \begin{bmatrix} 4\alpha_{11} & 2\alpha_{12} & \alpha_{13} \\ 4\alpha_{21} & 2\alpha_{22} & \alpha_{23} \\ 4\alpha_{31} & 2\alpha_{32} & \alpha_{33} \end{bmatrix} \begin{Bmatrix} x_1 \\ x_2 \\ x_3 \end{Bmatrix}$$

Now, $\alpha_{11} = \alpha_{12} = \alpha_{21} = \alpha_{13} = \alpha_{31} = \dfrac{1}{3k}$

$$\alpha_{22} = \alpha_{32} = \alpha_{23} = \frac{4}{3k}$$

$$\alpha_{33} = \frac{7}{3k}$$

Hence,

$$\begin{Bmatrix} x_1 \\ x_2 \\ x_3 \end{Bmatrix} = \frac{\omega^2 m}{3k} \begin{bmatrix} 4 & 2 & 1 \\ 4 & 8 & 4 \\ 4 & 8 & 7 \end{bmatrix} \begin{Bmatrix} x_1 \\ x_2 \\ x_3 \end{Bmatrix}$$

To start the iteration, a reasonable guess for the first mode is $\begin{Bmatrix} 1 \\ 2 \\ 4 \end{Bmatrix}$.

Eventually iteration for the first mode gives

$$\begin{Bmatrix} 1.0 \\ 3.2 \\ 4.0 \end{Bmatrix} = \frac{14.4\, m\omega^2}{3k} \begin{Bmatrix} 1.0 \\ 3.18 \\ 4.0 \end{Bmatrix}$$

or $\omega_1 = 0.46\sqrt{(k/m)}$ rad/s.

To obtain the second principal mode, use the orthogonality principle to remove the first mode from the system matrix;

$$m_1 A_1 A_2 + m_2 B_1 B_2 + m_3 C_1 C_2 = 0$$

Thus $4m\ (1{,}0)A_2 + 2m\ (3{,}18)B_2 + m(4{,}0)\ (C_2) = 0$

or $A_2 = -1.59\ B_2 - C_2$.

Hence $\begin{Bmatrix} A_2 \\ B_2 \\ C_2 \end{Bmatrix} = \begin{bmatrix} 0 & -1.6 & -1 \\ 0 & 1 & 0 \\ 0 & 0 & 1 \end{bmatrix} \begin{Bmatrix} A_2 \\ B_2 \\ C_2 \end{Bmatrix}$

When this sweeping matrix is combined with the original matrix equation, iteration makes convergence to the second mode take place.

$$\begin{Bmatrix} x_1 \\ x_2 \\ x_3 \end{Bmatrix} = \frac{\omega^2 m}{3k} \begin{bmatrix} 4 & 2 & 1 \\ 4 & 8 & 4 \\ 4 & 8 & 7 \end{bmatrix} \begin{bmatrix} 0 & -1.6 & -1 \\ 0 & 1 & 0 \\ 0 & 0 & 1 \end{bmatrix} \begin{Bmatrix} x_1 \\ x_2 \\ x_3 \end{Bmatrix}$$

$$= \frac{\omega^2 m}{3k} \begin{bmatrix} 0 & -4.4 & -3 \\ 0 & 1.6 & 0 \\ 0 & 1.6 & 3 \end{bmatrix} \begin{Bmatrix} x_1 \\ x_2 \\ x_3 \end{Bmatrix}$$

Give second mode as $\begin{Bmatrix} 1 \\ 0 \\ -1 \end{Bmatrix}$ and iterate:

$$\begin{Bmatrix} 1 \\ 0 \\ -1 \end{Bmatrix} = \frac{\omega^2 m}{3k} \begin{bmatrix} 0 & -4.4 & -3 \\ 0 & 1.6 & 0 \\ 0 & 1.6 & 3 \end{bmatrix} \begin{Bmatrix} 1 \\ 0 \\ -1 \end{Bmatrix} = \frac{m\omega^2}{k} \begin{Bmatrix} 1 \\ 0 \\ -1 \end{Bmatrix}$$

Hence $\omega_2 = \sqrt{(k/m)}$ rad/s.

To obtain the third mode, write the orthogonality principle as

$$m_1A_2A_3 + m_2B_2B_3 + m_3C_2C_3 = 0$$
$$m_1A_1A_3 + m_2B_1B_3 + m_3C_1C_3 = 0$$

Substitute $A_1 = 1.0\ B_1 = 3.18\ C_1 = 4.0$
$A_2 = 1.0\ B_2 = 0\quad C_2 = -1.0$

Hence $\begin{Bmatrix} A_3 \\ B_3 \\ C_3 \end{Bmatrix} = \begin{bmatrix} 0 & 0 & 0.25 \\ 0 & 0 & -0.78 \\ 0 & 0 & 1 \end{bmatrix} \begin{Bmatrix} A_3 \\ B_3 \\ C_3 \end{Bmatrix}$

When this sweeping matrix is combined with the equation for the second mode the second mode is removed, so that it yields the third mode on iteration;

$$\begin{Bmatrix} x_1 \\ x_2 \\ x_3 \end{Bmatrix} = \frac{\omega^2 m}{3k} \begin{bmatrix} 0 & -4.4 & -3 \\ 0 & 1.6 & 0 \\ 0 & 1.6 & 3 \end{bmatrix} \begin{bmatrix} 0 & 0 & 0.25 \\ 0 & 0 & -0.78 \\ 0 & 0 & 1 \end{bmatrix} \begin{Bmatrix} x_1 \\ x_2 \\ x_3 \end{Bmatrix}$$

$$= \frac{\omega^2 m}{3k} \begin{bmatrix} 0 & 0 & 0.43 \\ 0 & 0 & -1.25 \\ 0 & 0 & 1.75 \end{bmatrix} \begin{Bmatrix} x_1 \\ x_2 \\ x_3 \end{Bmatrix}$$

or $$\begin{Bmatrix} x_1 \\ x_2 \\ x_3 \end{Bmatrix} = 1.75 \frac{\omega^2 m}{3k} \begin{bmatrix} 0 & 0 & 0.25 \\ 0 & 0 & -0.72 \\ 0 & 0 & 1 \end{bmatrix} \begin{Bmatrix} x_1 \\ x_2 \\ x_3 \end{Bmatrix} .$$

Guess the third mode shape, iterate and find the third mode eigenvector as

$$\begin{Bmatrix} 0.25 \\ -0.72 \\ 1 \end{Bmatrix}$$

and $\omega_3 = 1.32\sqrt{(k/m)}$ rad/s.

The convergence for higher modes becomes more critical if impurities are introduced through the sweeping matrices.

It is well to check the highest mode by the inversion of the original matrix equation, which should be equal to the equation formulated in terms of the stiffness influence coefficients.

Example 14

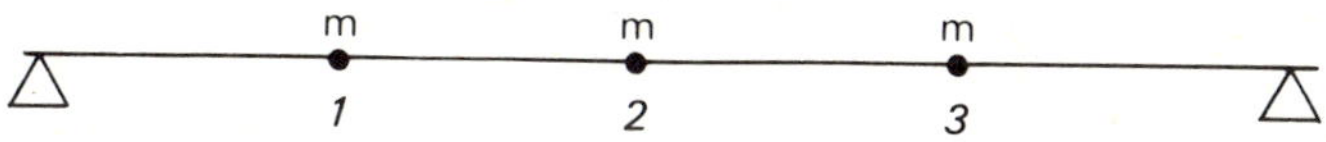

A simply supported beam of length L has three equal bodies each of mass m attached to it at equal distances as shown. Show that the influence coefficients are (where $\Delta = L^3/256\,EI$):

$$\begin{array}{lll} \alpha_{11} = 3\Delta & \alpha_{12} = 3.67\Delta & \alpha_{13} = 2.33\Delta \\ \alpha_{21} = 3.67\Delta & \alpha_{22} = 5.33\Delta & \alpha_{23} = 3.67\Delta \\ \alpha_{31} = 2.33\Delta & \alpha_{32} = 3.67\Delta & \alpha_{33} = 3\Delta \end{array}$$

Proceed to find the flexibility matrix and, by iteration, deduce the lowest natural frequency and associated mode shape.

3.3 THE LAGRANGE EQUATION

The Lagrange method is a very powerful method of analysis for certain physically complex systems. It is an energy method which allows the equations of motion to be written in terms of any set of generalised coordinates. Generalised coordinates are a set of independent parameters which completely specify the system location and which are independent of any constraints. The fundamental form of Lagrange's Equation can be written in terms of the generalised coordinates q_i as follows:

$$\frac{\mathrm{d}}{\mathrm{d}t}\left(\frac{\partial(\mathrm{KE})}{\partial \dot{q}_i}\right) - \frac{\partial(\mathrm{KE})}{\partial q_i} + \frac{\partial(\mathrm{PE})}{\partial q_i} + \frac{\partial(\mathrm{DE})}{\partial \dot{q}_i} = Q_i$$

where KE is the total kinetic energy of the system,
PE is the total potential energy of the system,
DE is the energy dissipation function, when the damping is linear. It is half the rate at which energy is dissipated, so that for viscous damping $\mathrm{DE} = \frac{1}{2} c\dot{x}^2$.
Q_i is a generalised external force (or a non-linear damping force) acting on the system and
q_i is a generalised coordinate that describes the position of the system.

The subscript i denotes n equations for an n degree of freedom system, so the Lagrange Equation yields as many equations of motion as there are degrees of freedom.

For a conservative system Q_i and DE are both zero so that:

$$\frac{\mathrm{d}}{\mathrm{d}t}\frac{\partial(\mathrm{KE})}{\partial \dot{q}_i} - \frac{\partial(\mathrm{KE})}{\partial q_i} + \frac{\partial(\mathrm{PE})}{\partial q_i} = 0 \quad .$$

The full derivation of the Lagrange Equation can be found in *Vibration Theory and Applications* by W. T. Thomson. Allen and Unwin. The application of the Lagrange Equation can best be demonstrated by means of an example.

Example 15

A solid cylinder has a mass M and radius R. Pinned to the axis of the cylinder is an arm of length l which carries a bob of mass m. Obtain the natural frequency of free vibration of the bob.

The generalised coordinates are x_1 and x_2. They completely specify the position of the system and are independent of any constraints.

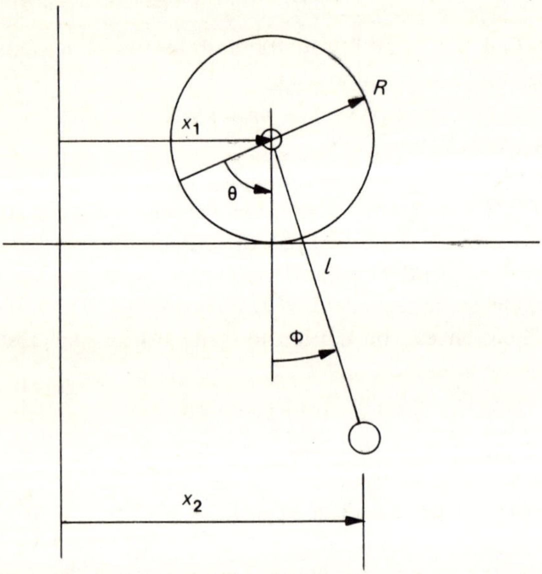

$$\mathrm{KE} = \tfrac{1}{2} M\dot{x}_1{}^2 + \tfrac{1}{2}\,(\tfrac{1}{2}\,MR^2)\dot{\theta}^2 + \tfrac{1}{2}\,m\dot{x}_2{}^2$$

$$= \tfrac{1}{2} M\dot{x}_1{}^2 + \tfrac{1}{2}\,(\tfrac{1}{2}\,M\dot{x}_1{}^2) + \tfrac{1}{2}\,m\dot{x}_2{}^2 \quad \text{, since } \dot{x} = R\dot{\theta}.$$

$$\mathrm{PE} = mgl(1-\cos\phi) = (mgl/2)\phi^2 = (mg/2l)\,(x_2 - x_1)^2 \quad .$$

Apply the Lagrange Equation with $q_i = x_1$:

$$(\mathrm{d}/\mathrm{d}t)\,(\partial(\mathrm{KE})/\partial\dot{x}_1) = M\ddot{x}_1 + \tfrac{1}{2}\,M\ddot{x}_1$$

$$\partial(\mathrm{PE})/\partial x_1 = (mg/2l)\,(-\,2x_2 + 2x_1) \quad .$$

Hence $\tfrac{3}{2}\,M\ddot{x}_1 + (mg/l)\,(x_1 - x_2) = 0$ is an equation of motion.

Apply the Lagrange Equation with $q_i = x_2$:

$$(\mathrm{d}/\mathrm{d}t)\,(\partial(\mathrm{KE})/\partial\dot{x}_2) = m\ddot{x}_2$$

$$\partial(\mathrm{PE})/\partial x_2 = (mg/2l)\,(2x_2 - 2x_1) \quad .$$

Hence $m\ddot{x}_2 + (mg/l)\,(x_2 - x_1) = 0$ is an equation of motion.

These equations of motion can be solved by assuming $x_1 = X_1 \sin\omega t$ and $x_2 = X_2 \sin\omega t$. Then

$$X_1\{(mg/l) - (3M/2)\,\omega^2\} + X_2\{-mg/l\} = 0$$

$$X_1\{-mg/l\} + X_2\{(mg/l) - m\omega^2\} = 0 \quad .$$

The frequency equation is therefore

$$(3M/2)\,\omega^4 - (g/l)\,\omega^2\{m + (3M/2)\} = 0 \quad .$$

Thus either $\omega = 0$ or $\omega = \sqrt{((1 + 2m/3M)\,g/l)}$ and $X_1/X_2 = -2m/3M$

Example 16
Find the equations of motion for free vibration of the triple pendulum shown.

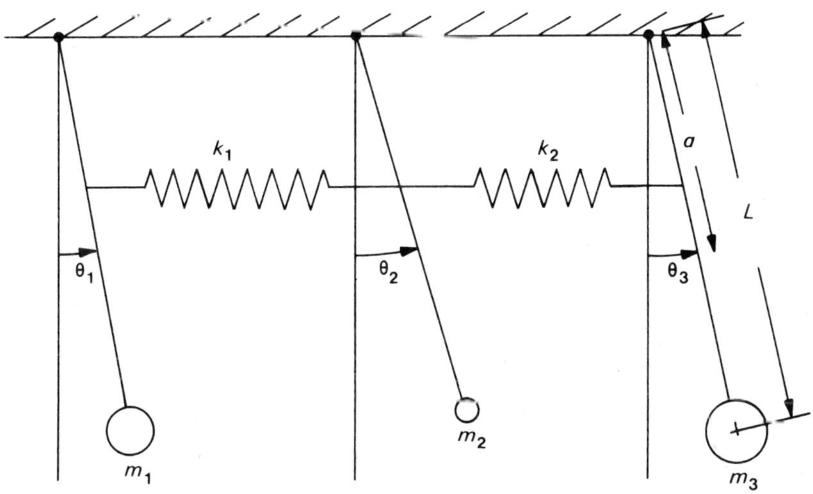

$$\text{KE} = \tfrac{1}{2}\, m_1 L^2 \dot{\theta}_1^{\,2} + \tfrac{1}{2}\, m_2 L^2 \dot{\theta}_2^{\,2} + \tfrac{1}{2}\, m_3 L^2 \dot{\theta}_3^{\,2}$$

$$\text{PE} = m_1 gL(1 - \cos\theta_1) + m_2 gL(1 - \cos\theta_2) + m_3 gL(1 - \cos\theta_3)$$

$$+ \tfrac{1}{2}\, k_1 (a\sin\theta_2 - a\sin\theta_1)^2 + \tfrac{1}{2}\, k_2 (a\sin\theta_3 - a\sin\theta_2)^2$$

Hence, by applying the Lagrange Equation with $q_i = \theta_1, \theta_2, \theta_3$ in turn, the equations of motion are obtained.

3.4 RECEPTANCES

Some simplification in the analysis of multi-degree of freedom undamped systems can often be gained by using receptances, particularly if only the natural frequencies are required. If an harmonic force $F\sin\upsilon t$ acts at some point in a system so that the system responds at frequency υ, and the point of application of the force has a displacement $x = X\sin\upsilon t$, then if the equations of motion are linear $x = \alpha F\sin\upsilon t$. α, which is a function of the system and υ but not a function of F, is known as the **direct receptance** at x. If the displacement is determined at some point other than at which the force is applied, α is known as the **transfer** or **cross receptance.**

It can be seen that the frequency at which a receptance becomes infinite is a natural frequency of the system. Receptances can be written for rotational and translational systems.

Thus, if a body of mass m is subjected to a force $F\sin\upsilon t$ and the response of the body is $x = X\sin\upsilon t$,

$$F\sin\upsilon t = m\ddot{x} = m(-X\upsilon^2 \sin\upsilon t) = -m\upsilon^2 x$$

i.e. $\quad x = -(1/m\upsilon^2)\, F\sin\upsilon t$

and $\quad \alpha = -1/m\upsilon^2$. This is the direct receptance of a rigid body.

For a spring, $\alpha = 1/k$. This is the direct receptance of a spring.

In an undamped single degree of freedom system,

$$m\ddot{x} + kx = F\sin\upsilon t$$

If $x = X\sin\upsilon t$, $\alpha = 1/(k - m\upsilon^2)$. This is the direct receptance of a single degree of freedom system.

In more complicated systems, it is necessary to be able to distinguish between direct and cross receptances and to specify the points at which the receptances are calculated. This is done by using subscripts. The first subscript indicates the coordinate at which the response is measured, and the second indicates that at which the force is applied. Thus α_{pq} which is a cross receptance is the response at p divided by the harmonic force applied at q, and α_{pp} and α_{qq} are direct receptances at p and q respectively.

Consider the two degree of freedom system shown in Fig. 3.12.

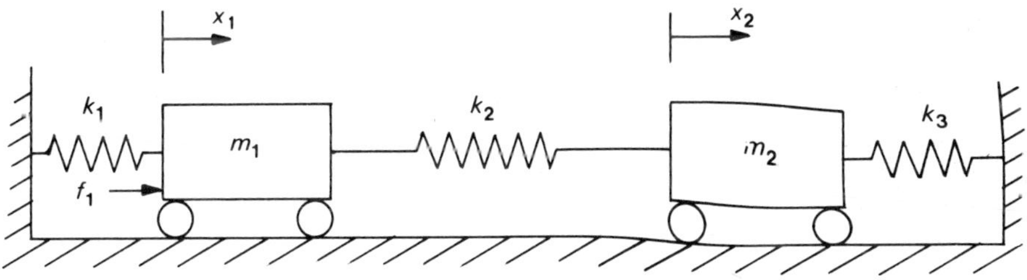

Fig. 3.12

The equations of motion are

$$m_1\ddot{x}_1 + (k_1 + k_2)x_1 - k_2x_2 = f_1$$

$$m_2\ddot{x}_2 + (k_2 + k_3)x_2 - k_2x_1 = 0 \quad .$$

Let $f_1 = F_1 \sin \upsilon t$ and assume $x_1 = X_1 \sin \upsilon t$ and $x_2 = X_2 \sin \upsilon t$. Substitute into the equations of motion:

$$(k_1 + k_2 - m_1\upsilon^2)X_1 + (-k_2)X_2 = F_1$$

$$(-k_2)X_1 + (k_2 + k_3 - m_2\upsilon^2)X_2 = 0 \quad .$$

Thus $\alpha_{11} = X_1/F_1 = (k_2 + k_3 - m_2\upsilon^2)/\Delta$

where $\Delta = (k_1 + k_2 - m_1\upsilon^2)\,(k_2 + k_3 - m_2\upsilon^2) - k_2^{\,2}$

and $\Delta = 0$ is the frequency equation.

α_{11} is a direct receptance

$$\alpha_{21} = X_2/F_1 = k_2/\Delta$$

α_{21} is a cross receptance.

This system has two more receptances, the responses due to f_2 applied to the second body. Thus α_{12} and α_{22} may be found. It is a fundamental property that $\alpha_{12} \equiv \alpha_{21}$, (Principle of Reciprocity) so that symmetrical matrices result.

A general statement of the system response is

$$X_1 = \alpha_{11} F_1 + \alpha_{12} F_2$$

and $$X_2 = \alpha_{21} F_1 + \alpha_{22} F_2$$

i.e. $$\begin{Bmatrix} X_1 \\ X_2 \end{Bmatrix} = \begin{bmatrix} \alpha_{11} & \alpha_{12} \\ \alpha_{21} & \alpha_{22} \end{bmatrix} \begin{Bmatrix} F_1 \\ F_2 \end{Bmatrix} .$$

Some simplification in the analysis of complex systems can be achieved by considering the complex system to be a number of simple systems (whose receptances are known) linked together by using conditions of compatibility and equilibrium. The method is to break the complex system down into sub-systems and analyse each sub-system separately. Find each sub-system receptance at the point where it is connected to the adjacent sub-system, and 'join' all sub-systems together using the conditions of compatibility and equilibrium.

For example, to find the direct receptance γ_{11} of a system C at a single coordinate x_1 the system is considered as two sub-systems A and B, Fig. 3.13.

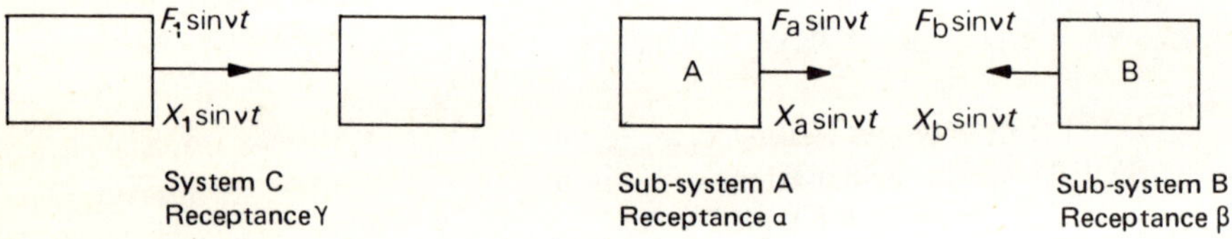

Fig. 3.13

By definition $\gamma_{11} = X_1/F_1$, $\alpha_{11} = X_a/F_a$ and $\beta_{11} = X_b/X_b$.

Because the systems are connected,

$$X_a = X_b = X_1 \qquad \text{(Compatibility)}$$

and $$F_1 = F_a + F_b \ . \qquad \text{(Equilibrium)}$$

Hence $1/\gamma_{11} = (1/\alpha_{11}) + (1/\beta_{11})$ i.e. the system receptance γ can be found from the receptances of the sub-systems.

In a simple spring-body system, sub-systems A and B are the spring and body respectively. Hence $\alpha_{11} = 1/k$ and $\beta_{11} = -1/m\nu^2$ and $1/\gamma_{11} = k - m\nu^2$, as above.

The frequency equation is $\alpha_{11} + \beta_{11} = 0$, because this condition makes $\gamma_{11} = \infty$.

Consider the undamped dynamic vibration absorber application shown in Fig. 3.14. The system is split into sub-systems A and B.

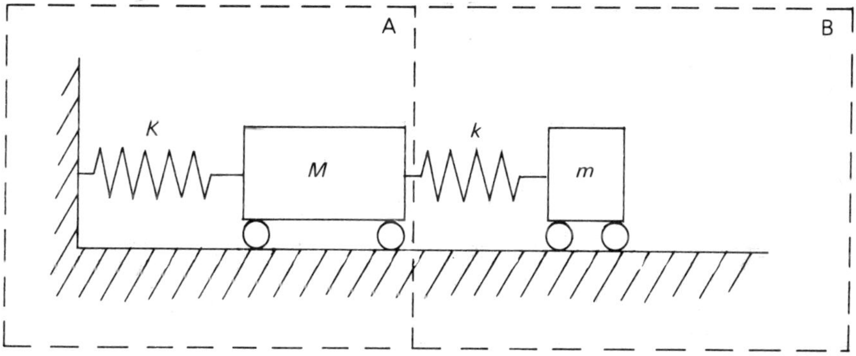

Fig. 3.14

For sub-system A,

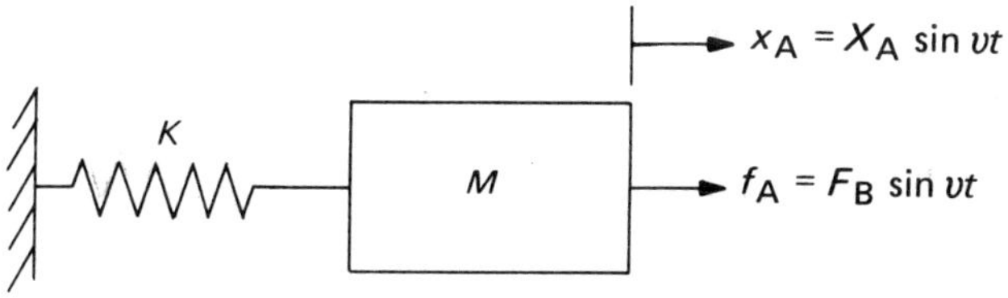

$$f_A = M\ddot{x}_A + Kx_A; \ \alpha = 1/(K - M\upsilon^2).$$

For sub-system B,

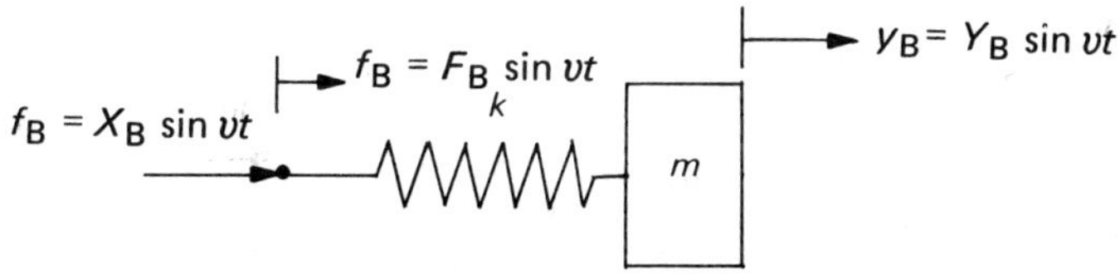

$$f_B = k(x_B - y_B) = m\ddot{y}_B = -m\upsilon^2 Y_B, \ \beta = -(k - m\upsilon^2)/km\upsilon^2 .$$

Thus the frequency equation $\alpha + \beta = 0$ gives

$$Mm\upsilon^4 - (mK + Mk + mk)\upsilon^2 + Kk = 0 \quad .$$

It is often convenient to solve the frequency equation $\alpha + \beta = 0$, or $\alpha = -\beta$ by a graphical method. In the case of the absorber, both α and $-\beta$ can be plotted as a function of v, and the intersections give the natural frequencies Ω_1 and Ω_2, Fig. 3.15.

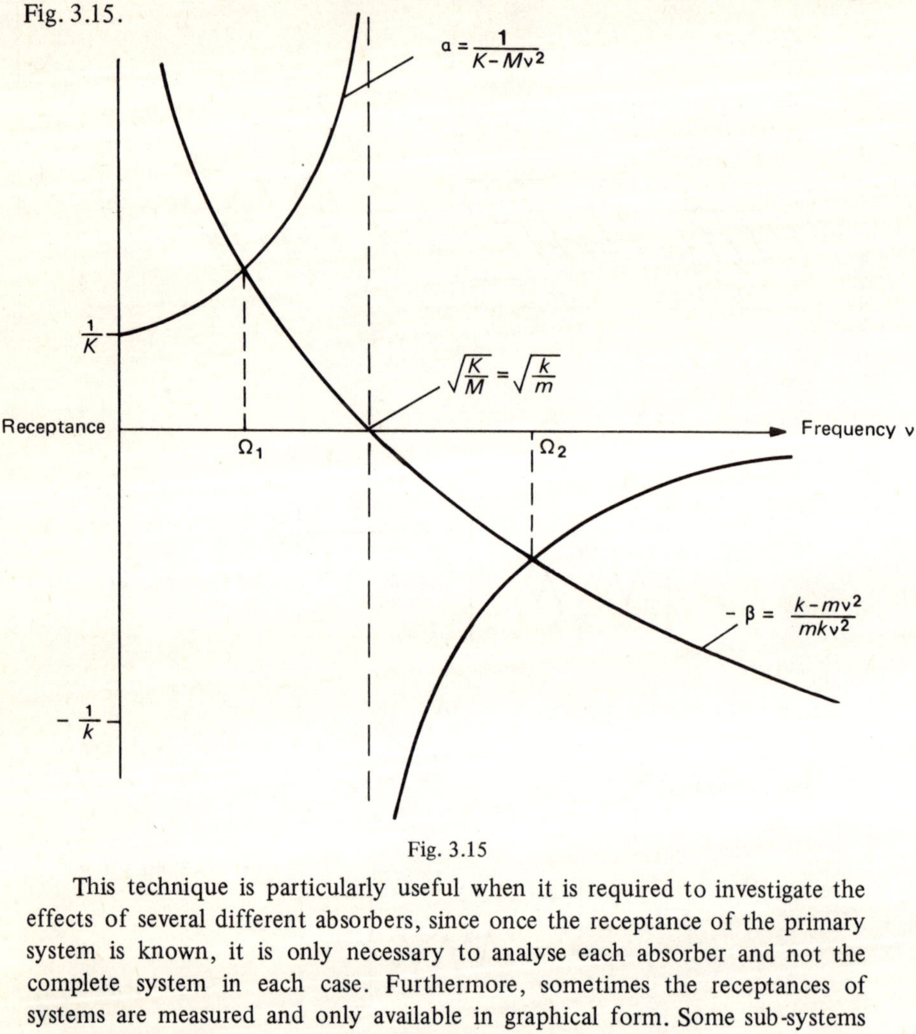

Fig. 3.15

This technique is particularly useful when it is required to investigate the effects of several different absorbers, since once the receptance of the primary system is known, it is only necessary to analyse each absorber and not the complete system in each case. Furthermore, sometimes the receptances of systems are measured and only available in graphical form. Some sub-systems such as those shown in Fig. 3.16 are linked by two coordinates, for example deflection and slope at the common point.

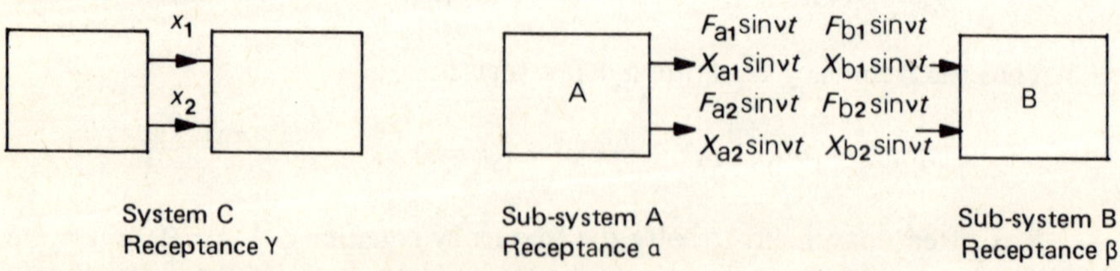

Fig. 3.16

Now

$$X_{a1} = \alpha_{11}F_{a1} + \alpha_{12}F_{a2}, \ X_{a2} = \alpha_{21}F_{a1} + \alpha_{22}F_{a2}$$

and

$$X_{b1} = \beta_{11}F_{b1} + \beta_{12}F_{b2}, \ X_{b2} = \beta_{21}F_{b1} + \beta_{22}F_{b2} \quad .$$

The applied forces are $F_1 \sin \nu t$ and $F_2 \sin \nu t$ where

$$F_1 = F_{a1} + F_{b1}, \ F_2 = F_{a2} + F_{b2}$$

Since the sub-systems are linked

$$X_1 = X_{a1} = X_{b1}, \ X_2 = X_{a2} = X_{b2}$$

Hence, if excitation is applied at x_1 only, $F_2 = 0$ and

$$\gamma_{11} = \frac{X_1}{F_1} = \frac{\alpha_{11}(\beta_{11}\beta_{22} - \beta_{12}^2) + \beta_{11}(\alpha_{11}\alpha_{22} - \alpha_{12}^2)}{\Delta}$$

where $\Delta = (\alpha_{11} + \beta_{11})(\alpha_{22} + \beta_{22}) - (\alpha_{12} + \beta_{12})^2$

$$\gamma_{21} = \frac{X_2}{F_1} = \frac{\alpha_{12}(\beta_{11}\beta_{22} - \alpha_{12}\beta_{12}) - \beta_{12}(\alpha_{11}\alpha_{22} - \alpha_{12}\beta_{12})}{\Delta}$$

If $F_1 = 0$

$$\gamma_{22} = \frac{X_2}{F_2} = \frac{\alpha_{22}(\beta_{11}\beta_{22} - \beta_{12}^2) - \beta_{22}(\alpha_{11}\alpha_{22} - \alpha_{12}^2)}{\Delta}$$

Since $\Delta = 0$ is the frequency equation the natural frequencies of the system C are given by

$$\begin{vmatrix} \alpha_{11} + \beta_{11} & \alpha_{12} + \beta_{12} \\ \alpha_{21} + \beta_{21} & \alpha_{22} + \beta_{22} \end{vmatrix} = 0$$

This is an extremely useful method for finding the frequency equation of a system because only the receptances of the sub-systems are required. The receptances of many systems have been published by R. E. D. Biship and D. C. Johnson in *The Mechanics of Vibration,* CUP. By repeated application of this method, a system can be considered to comprise any number of sub-systems.

It should be appreciated that although the receptance technique is useful for writing the frequency equation, it does not simplify the solution of this equation.

CHAPTER 4

The vibration of systems with distributed mass and elasticity

Continuous systems such as beams, rods, cables and strings can be modelled by discrete mass and stiffness parameters and analysed as multi degree of freedom systems, but such a model is not sufficiently accurate for most purposes. Furthermore, mass and elasticity cannot always be separated in models of real systems. Thus mass and elasticity have to be considered as distributed parameters.

For the analysis of systems with distributed mass and elasticity it is necessary to assume an homogeneous, isotropic material which follows Hooke's Law.

Generally free vibration is the sum of the principal modes. However, in the unlikely event of the elastic curve of the body in which motion is excited coinciding exactly with one of the principal modes, only that mode will be excited. In most continuous systems the rapid damping out of high-frequency modes often leads to the fundamental mode predominating.

4.1 WAVE MOTION

4.1.1 Transverse vibration of a string

Consider a vibrating flexible string of mass ρ per unit length and stretched under a tension T as shown in Fig. 4.1.

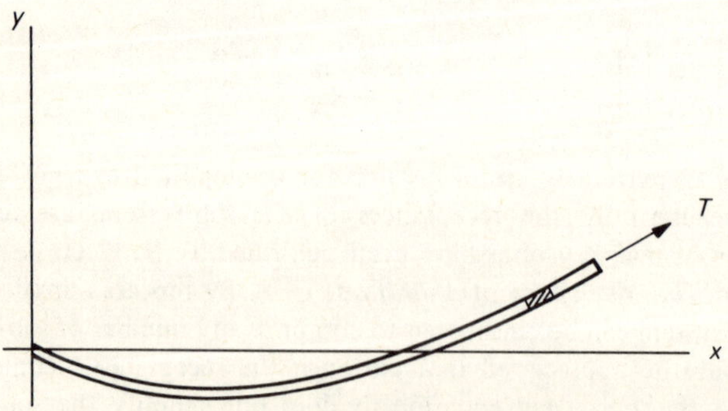

Fig. 4.1

The FBDs of an element of length dx of the string assuming small deflections and slopes are given in Fig. 4.2. By assuming the lateral deflection y to be small, the change in tension with deflection is negligible, and $\sin\theta \simeq \theta$.

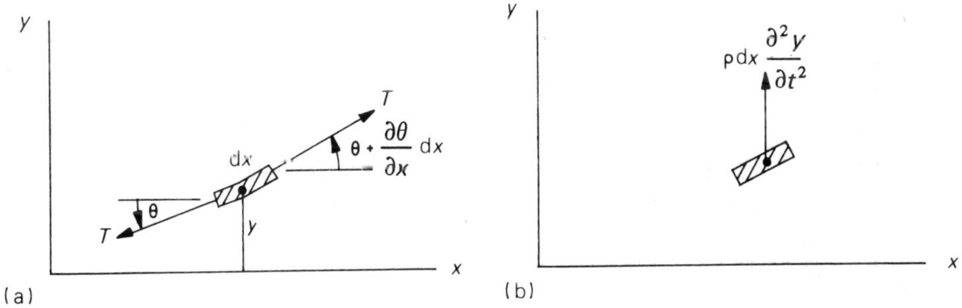

Fig. 4.2(a) – Applied forces. (b) Effective force.

The equation for motion in the y direction is

$$T(\theta + (\partial\theta/\partial x)\,\mathrm{d}x) - T\,\theta = \rho\,\mathrm{d}x\,(\partial^2 y/\partial t^2)$$

so $\quad \partial\theta/\partial x = (\rho/T)\,(\partial^2 y/\partial t^2)$

since $\quad \theta = \partial y/\partial x$,

$$\partial^2 y/\partial x^2 = (1/c^2)\,(\partial^2 y/\partial t^2) \qquad \text{where } c = \sqrt{(T/\rho)}. \tag{4.1}$$

This is the Wave Equation, the solution of which is given in section 4.1.4. The velocity of propagation of waves along the string is c.

4.1.2 Longitudinal vibration of a thin uniform bar

Consider the longitudinal vibration of a thin uniform bar of cross-sectional area S, density ρ and modulus E under an axial force P, as shown in Fig. 4.3.

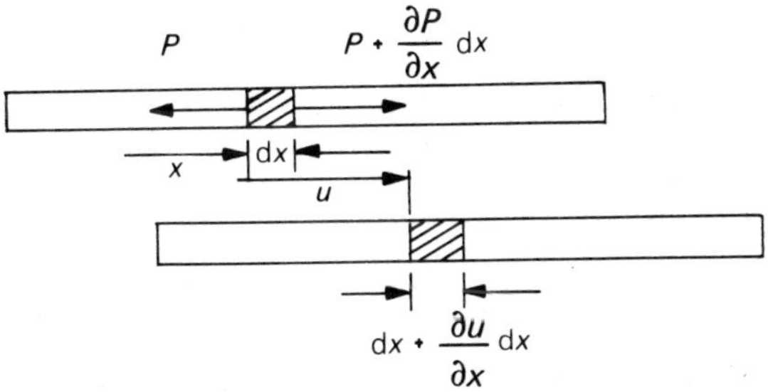

Fig. 4.3

From Fig. 4.3 $(\partial P/\partial x)\,dx = \rho\, S dx(\partial^2 u/\partial t^2)$

Now strain $\partial u/\partial x = P/SE$ so $\partial P/\partial x = SE(\partial^2 u/\partial x^2)$

Thus $\partial^2 u/\partial t^2 = (E/\rho)\,(\partial^2 u/\partial x^2)$

or $\partial^2 u/\partial x^2 = (1/c^2)\,(\partial^2 u/\partial t^2)$, where $c = \sqrt{(E/\rho)}$. (4.2)

This is the Wave Equation. The velocity of propagation of the displacement or stress wave in the rod is c.

4.1.3 Torsional vibration of a uniform shaft

In the uniform shaft shown in Fig. 4.4 the modulus of rigidity in torsion is G, the material density is ρ, and the polar second moment of area about the axis of twist is J.

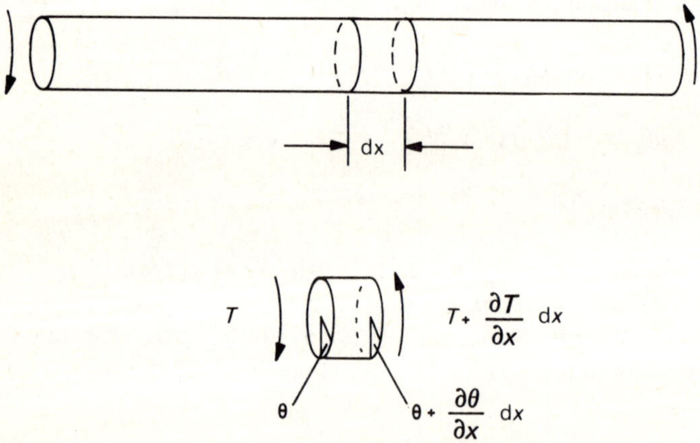

Fig. 4.4

From Fig. 4.4, $(\partial T/\partial x)\,dx = \rho\ J\ dx(\partial^2\theta/\partial t^2)$.

The elastic equation is $T = GJ(\partial\theta/\partial x)$

Thus $\partial^2\theta/\partial t^2 = (G/\rho)\,(\partial^2\theta/\partial x^2)$

or $\partial^2\theta/\partial x^2 = (1/c^2)\,(\partial^2\theta/\partial t^2)$, where $c = \sqrt{(G/\rho)}$. (4.3)

This is the Wave Equation. The velocity of propagation of the shear stress wave in the shaft is c.

4.1.4 Solution of the Wave Equation

The Wave Equation

$$\partial^2 y/\partial x^2 = (1/c^2)\,(\partial^2 y/\partial t^2)$$

can be solved by the method of separation of variables and assuming a solution of the form

$$y(x, t) = U(x)\,G(t).$$

Substituting this solution into the wave equation gives:

$$(\partial^2 U/\partial x^2)G(t) = (1/c^2)\,(\partial^2 G/\partial t^2)\,U(x)$$

i.e. $1/U(x)\,.\,(\partial^2 U/\mathrm{d}x^2) = (1/c^2)\,(1/G(t))\,(\mathrm{d}^2 G/\mathrm{d}t^2).$

The LHS is a function of x only, and the RHS is a function of t only, so partial derivatives are no longer required. Each side must be a constant, $-(\omega/c)^2$ say. (This quantity is chosen for convenience of solution).

Then $(\mathrm{d}^2 U/\mathrm{d}x^2) + (\omega/c)^2\,U = 0$

and $(\mathrm{d}^2 G/\mathrm{d}t^2) + \omega^2\,G = 0$.

Hence $U(x) = A\,\sin(\omega/c)x + B\,\cos(\omega/c)x$

and $G(t) = C\,\sin\omega t + D\,\cos\omega t$.

The constants A and B depend on the boundary conditions, and C and D on the initial conditions. The complete solution to the Wave Equation is therefore:

$$u = (A\,\sin(\omega/c)x + B\,\cos(\omega/c)x)\,(C\,\sin\omega t + D\,\cos\omega t) \quad . \qquad (4.4)$$

Example 17

Find the natural frequencies and mode shapes of longitudinal vibrations for a free-free beam with initial displacement zero.

Since free ends, $\partial u/\partial x = 0$ at $x = 0$ and $x = l$.

$$\partial u/\partial x = (A(\omega/c)\cos(\omega/c)\,x - B\,(\omega/c)\sin(\omega/c)\,x)\,(C\,\sin\omega t + D\,\cos\omega t) \ .$$

$$(\partial u/\partial x)_{x=0} = A(\omega/c)\,(C\,\sin\omega t + D\,\cos\omega t) = 0, \text{ therefore } A = 0 \ .$$

$$(\partial u/\partial x)_{x=l} = (\omega/c)\,(-B\,\sin(\omega l/c))\,(C\,\sin\omega t + D\,\cos\omega t) = 0 \ .$$

Thus $\sin(\omega l/c) = 0$,

and $\omega l/c = \omega l/\sqrt{(E/\rho)} = \pi, 2\pi, \ldots n\pi$.

$\omega_n = (n\pi/l)\sqrt{(E/\rho)}$ where $\omega = c/\lambda$ and λ is the wavelength.

If initial displacement is zero, $D = 0$ and $u = B\cos(n\pi/l)x \,.\, \sin(n\pi/l)\sqrt{(E/\rho)}t$.

Example 18

A uniform vertical rod of length l and cross-section S is fixed at the upper end and is loaded with a body of mass M on the other. Show that the natural frequencies of longitudinal vibration are determined by:

$$\omega l\sqrt{(\rho/E)}\tan\omega l\sqrt{(\rho/E)} = S\rho l/M$$

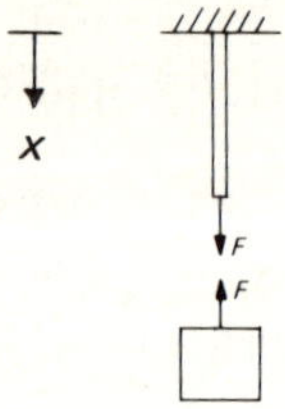

At $x = 0$ $u = 0$

and at $x = l$, $F = SE\,(\partial u/\partial x)$

Also $F = SE\,(\partial u/\partial x) = -M(\partial^2 u/\partial t^2)$

General solution is $u = (A\sin(\omega/c)x + B\cos(\omega/c)x)(C\sin\omega t + D\cos\omega t)$

Now, $u_{x=0} = 0$ so $B = 0$

thus $u = (A\sin(\omega/c)x)(C\sin\omega t + D\cos\omega t)$

$$(\partial u/\partial x)_{x=l} = (A(\omega/c)\cos(\omega l/c))(C\sin\omega t + D\cos\omega t)$$

and $(\partial^2 u/\partial t^2)_{x=l} = (-A\omega^2\sin(\omega l/c))(C\sin\omega t + D\cos\omega t)$

so, $F = SE\,(\omega/c)\cos(\omega l/c)(C\sin\omega t + D\cos\omega t) = M\omega^2\sin(\omega l/c)(C\sin\omega t + D\cos\omega t)$.

Hence $(\omega l/c)\tan(\omega l/c) = SlE/Mc^2$

and $\omega l\sqrt{(\rho/E)}\tan\omega l\sqrt{(\rho/E)} = S\rho l/M$ since $c^2 = E/\rho$.

4.2 TRANSVERSE VIBRATION

4.2.1 Transverse vibration of a uniform beam

The transverse or lateral vibration of a thin uniform beam is another vibration problem in which both elasticity and mass are distributed. Consider the moments and forces acting on the element of the beam shown in Fig. 4.5. The beam has a cross-sectional area A, flexural rigidity EI, and material of density ρ.

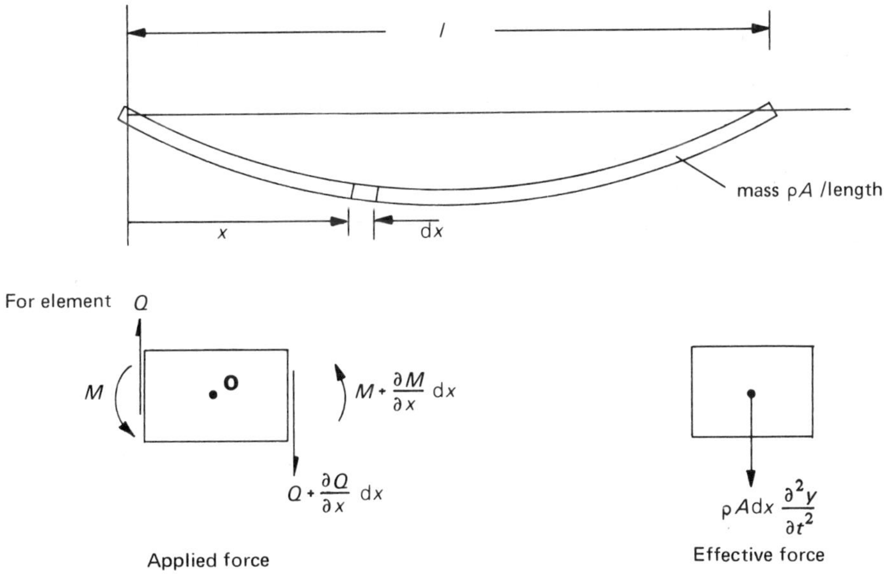

Fig. 4.5

Then for the element, neglecting rotary inertia and shear of the element and taking moments about O, gives:

$$M + Q\,(\mathrm{d}x/2) + Q\,(\mathrm{d}x/2) + (\partial Q/\partial x)\,\mathrm{d}x\,(\mathrm{d}x/2) = M + (\partial M/\partial x)\,\mathrm{d}x,$$

i.e. $Q - \partial M/\partial x$ (4.5)

Summing forces in the y direction gives:

$$(\partial Q/\partial x)\,\mathrm{d}x = \rho A\,\mathrm{d}x\,(\partial^2 y/\partial t^2)\quad .$$

Hence $\partial^2 M/\partial x^2 = \rho A\,(\partial^2 y/\partial t^2)$.

Now EI is a constant for a prismatical beam, so

$$M = -EI\,(\partial^2 y/\partial x^2) \text{ and } \partial^2 M/\partial x^2 = -EI\,(\partial^4 y/\partial x^4)\quad .$$

Thus $(\partial^4 y/\partial x^4) + (\rho A/EI)\,(\partial^2 y/\partial t^2) = 0$. (4.6)

This is the general equation for the transverse vibration of a uniform beam.

When a beam performs a normal mode of vibration the deflection at any point of the beam varies harmonically with time and can be written

$$y = X(B_1 \sin\omega t + B_2 \cos\omega t)$$

where X is a $f(x)$ which defines the beam shape of the normal mode of vibration.

$$\text{Hence } \mathrm{d}^4X/\mathrm{d}x^4 = (\rho A/EI)\,\omega^2 X = \lambda^4 X \tag{4.7}$$

where $\lambda^4 = \rho A\omega^2/EI$.

The general solution to equation (4.7) is

$$X = C_1 \cos\lambda x + C_2 \sin\lambda x + C_3 \cosh\lambda x + C_4 \sinh\lambda x \quad . \tag{4.8}$$

where the constants $C_{1,\,2,\,3,\,4}$ are determined from the boundary conditions.

For example, consider the transverse vibration of a thin prismatical beam simply supported at each end.

At $x = 0$ and $x = l$, $X = 0$ and $\mathrm{d}^2X/\mathrm{d}x^2 = 0$.

Thus $C_1 = C_3 = 0$ and $X = C_2 \sin\lambda x + C_4 \sinh\lambda x$

and $C_2 \sin\lambda l = C_4 \sinh\lambda l = 0$.

Since $\lambda l \neq 0$ $\sinh\lambda l \neq 0$ and therefore $C_4 = 0$,

and $C_2 \sin\lambda l = 0$, so $C_2 \neq 0$, therefore $\sin\lambda l = 0$.

Hence $X = C_2 \sin\lambda x$

The solutions to $\sin\lambda l = 0$ give the natural frequencies,

i.e. $\lambda = 0, \pi/l, 2\pi/l, 3\pi/l \ldots$.

and $\omega = 0, (\pi/l)^2\sqrt{(EI/A\rho)}, (2\pi/l)^2\sqrt{(EI/A\rho)}, (3\pi/l)^2\sqrt{(EI/A\rho)} \ldots$

$\lambda = 0$, $\omega = 0$ is a trivial solution because the beam is at rest. The lowest or first natural frequency is therefore $\omega_1 = (\pi/l)^2\sqrt{(EI/A\rho)}$ rad/s and the corresponding mode shape is $X = C_2 \sin\pi\,(x/l)$. This is the first mode. $\omega_2 = (2\pi/l)^2\sqrt{(EI/A\rho)}$ rad/s is the second natural frequency and the second mode is $X = C_2 \sin 2\pi x/l$, and so on. The mode shapes are drawn in Fig. 4.6.

1st mode shape, one half wave:

$$y = C_2 \sin(\pi x/l)\,(B_1 \sin\omega_1 t + B_2 \cos\omega_1 t);\;\; \omega_1 = (\pi/l)^2\sqrt{(EI/A\rho)}\ \text{rad/s}$$

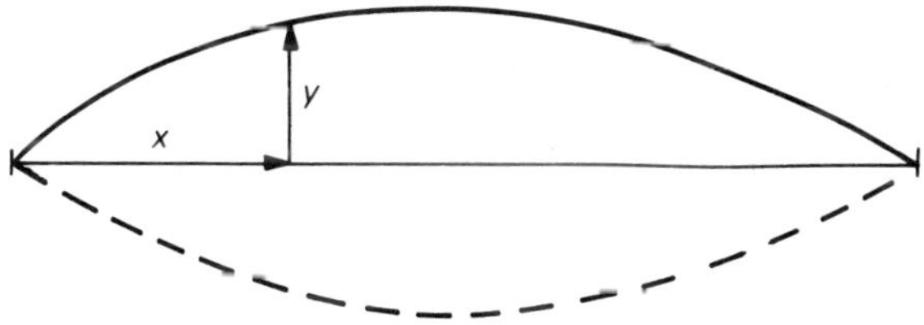

2nd mode shape, two half waves:

$$y = C_2 \sin(2\pi x/l)\,(B_1 \sin\omega_2 t + B_2 \cos\omega_2 t);\;\; \omega_2 = (2\pi/l)^2\sqrt{(EI/A\rho)}\ \text{rad/s}$$

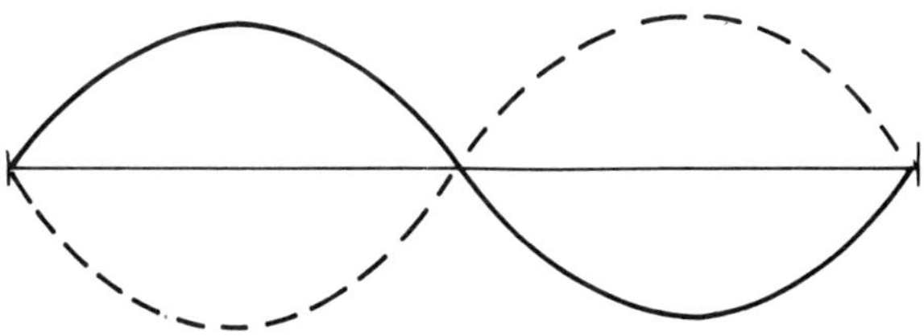

3rd mode shape, three half waves:

$$y = C_2 \sin(3\pi x/l)\,(B_1 \sin\omega_3 t + B_2 \cos\omega_3 t);\;\; \omega_3 = (3\pi/l)^2\sqrt{(EI/A\rho)}\ \text{rad/s}$$

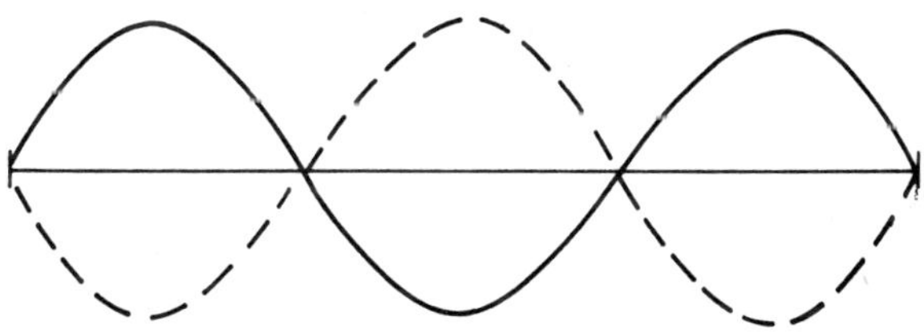

Fig. 4.6

These sinusoidal vibrations can be superimposed so that any initial conditions can be represented.

Other end conditions give frequency equations with the solution

$$\omega = (\alpha/l^2)\sqrt{(EI/A\rho)} \qquad \text{rad/s.}$$

where the values of α are given in the following table.

Table 4.1

Frequency equations and their solution for the first five modes of transverse vibration of beams. Solutions that is the natural frequencies are:

$$\omega = \frac{\alpha}{l^2}\sqrt{\frac{EI}{A\rho}} \text{ rad/s}$$

where α is given below.

End conditions	Frequency equation	1st mode	2nd mode	3rd mode	4th mode	5th mode
Clamped-free	$\cos\lambda l \cosh\lambda l = -1$	3.52	22.4	61.7	121.0	199.9
Pinned-pinned	$\sin\lambda l = 0$	9.87	39.5	88.9	157.9	246.8
Clamped-pinned	$\tan\lambda l = \tanh\lambda l$	15.4	50.0	104.0	178.3	272.0
Clamped-clamped or free-free	$\cos\lambda l \cosh\lambda l = 1$	22.4	61.7	121.0	199.9	298.6

4.2.2 The whirling of shafts

An important application of the theory for transverse beam vibration is to the whirling of shafts. If the speed of rotation of a shaft is increased, certain speeds will be reached at which violent instability occurs. These are the critical speeds of whirling. Since the loading on the shaft is due to centrifugal effects the equation of motion is exactly the same as for transverse beam vibration. The centrifugal effects occur because it is impossible to make the centre of mass of any section coincide exactly with the axis of rotation, because of a lack of homogeneity in the material and other practical difficulties.

4.2.3 Rotary inertia and shear effects

When a beam is subjected to lateral vibration so that the depth of the beam is a significant proportion of the distance between two adjacent modes, rotary inertia and transverse shear deformation make significant contributions to the lateral deflection. Thus rotary inertia and shear effects must be taken into account in the analysis of high-frequency vibration of all beams and in all analyses of deep beams.

Equation (4.5) can be modified to take into account rotary inertia by adding a term

$\rho I(\partial^3 y/(\partial x \partial t^2))$ so that equation (4.6) becomes

$$EI(\partial^4 y/\partial x^4) - \rho I(\partial^4 y/(\partial x^2 \partial t^2)) + \rho A(\partial^2 y/\partial t^2) = 0 \quad .$$

Shear deformation effects can be included by adding a term $(EI\rho/kG)$ $(\partial^4 y/\partial x^2 \partial t^2)$ where k is a constant whose value depends on the cross-section of the beam. Generally k is about 0.85. Equation (4.6) then becomes

$$EI(\partial^4 y/\partial x^4) - (EI\rho/kG)\,(\partial^4 y/(\partial x^2 \partial t^2)) + \rho A(\partial^2 y/\partial t^2) = 0 \quad .$$

Solutions to these equations are available, which generally lead to a frequency a few per cent more accurate than the solution to equation (4.6). However, in most cases the modelling errors exceed this. In general, the correction due to shear is larger than the correction due to rotary inertia.

Example 19

Find the first three natural frequencies of a steel bar 3 cm diameter which is simply supported at each end and has a length 1.5 m. Take $\rho = 7780$ kg/m^3 and $E = 208$ GN/m^2. What should the length be reduced to in order to raise the first natural frequency to 100 Hz?

$$\sqrt{(EI/A\rho)} = \sqrt{(208.10^9 \pi (0.03)^4 / 64 \pi (0.03/2)^2 7780)} \text{ m/s}^2$$

$$= 38.8 \text{ m/s}^2$$

$$\omega_1 = (\pi^2/1.5^2)\,38.8 = 170.2 \text{ rad/s, i.e. } f_1 = 27.1 \text{ Hz}$$

$$f_2 = 27.1 \times 4 = 108.4 \text{ Hz} \quad .$$

$$f_3 = 27.1 \times 9 = 243.9 \text{ Hz} \quad .$$

$$l_2 = l_1 \sqrt{(f_1/f_2)}$$

$$l_2 = 1.5 \sqrt{(27.1/100)} = 0.78 \text{ m} \quad .$$

4.2.4 Transverse vibration of a beam with discrete bodies

In those cases where it is required to find the lowest frequency of transverse vibration of a beam which carries discrete bodies, or the lowest critical speed of a shaft which carries discrete rotors, Dunkerley's method may be used.

Dunkerley's method uses the following equation:

$$1/\omega_1^2 \simeq (1/P_1^2) + (1/P_2^2) + (1/P_3^2) + \ldots$$

where ω_1 is the lowest natural frequency of a system and P_1, P_2, P_3 etc are the frequencies of each body acting alone.

This equation may be obtained for a two degree of freedom system by writing the equations of motion in terms of the influence coefficients as follows:

$$y_1 = \alpha_{11} m_1 \omega^2 y_1 + \alpha_{12} m_2 \omega^2 y_2$$

$$y_2 = \alpha_{21} m_1 \omega^2 y_1 + \alpha_{22} m_2 \omega^2 y_2 \quad .$$

The frequency equation is given by;

$$\begin{vmatrix} \alpha_{11} m_1 \omega^2 - 1 & \alpha_{12} m_2 \omega^2 \\ \alpha_{21} m_1 \omega^2 & \alpha_{22} m_2 \omega^2 - 1 \end{vmatrix} = 0$$

By expanding this determinant and solving the resulting quadratic equation it is found that;

$$\omega_{1,2}^2 = \frac{(\alpha_{11} m_1 + \alpha_{22} m_2) \pm \sqrt{((\alpha_{11} m_1 + \alpha_{22} m_2)^2 - 4(\alpha_{11}\alpha_{22} - \alpha_{21}\alpha_{12}))}}{2(\alpha_{11}\alpha_{22} - \alpha_{21}\alpha_{12})}$$

Hence it can be shown that;

$$(1/\omega_1^2) + (1/\omega_2^2) = \alpha_{11} m_1 + \alpha_{22} m_2 \quad .$$

Now P_1 is the natural frequency of body 1 acting alone,

hence $P_1^2 = k_1/m_1 = 1/(\alpha_{11}m_1)$. Similarly $P_2^2 = 1/(\alpha_{22}m_2)$.

Thus $(1/\omega_1^2) + (1/\omega_2^2) = (1/P_1^2) + (1/P_2^2)$.

A similar relationship can be derived for systems with more than two degrees of freedom.

If $\omega_2 >> \omega_1$, the left-hand side is approximately $1/\omega_1^2$,

hence $1/\omega_1^2 \simeq (1/P_1^2) + (1/P_2^2)$.

Example 20

A steel shaft (ρ = 8000 kg/m^3 E = 210 GN/m^2) 0.055 m diameter, running in self-aligning bearings 1.25 m apart, carries a rotor of mass 70 kg, 0.4 m from one bearing. Estimate the lowest critical speed.

For the shaft alone

$$P_1 = (\pi/l)^2 \sqrt{(EI/A\rho)} \text{ rad/s} = 141 \text{ rad/s} = 1350 \text{ rev/min.}$$

This is the lowest critical speed for the shaft without the rotor. For the rotor alone, neglecting the mass of the shaft,

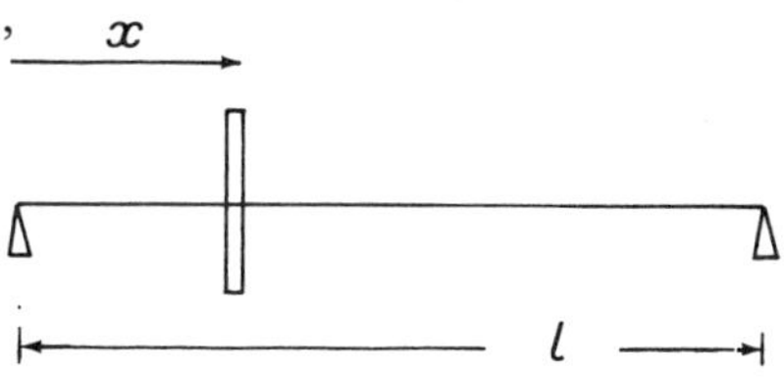

$$P_2 = \sqrt{(k/m)} \text{ rad/s}$$

$$k = 3EIl/(x^2(L-x)^2)$$

where x = 0.4 m and L = 1.25 m

Thus k = 3.07 MN/m

and P_2 = 209 rad/s = 2000 rev/min.

Now

$1/N_1^2 = 1/1350^2 + 1/2000^2$, hence N_1 = 1135 rev/min

4.3 THE ANALYSIS OF CONTINUOUS SYSTEMS BY RAYLEIGH'S METHOD

Energy methods can be used in the analysis of continuous systems if the mode shape of the system is known; this is necessary for evaluating the kinetic and potential energies. It has been shown by Rayleigh that the lowest natural frequency of such systems can be fairly accurately found by assuming any reasonable deflection curve for the shape of the first mode. Naturally the curve must be compatible with the end conditions of the system, and since any deviation from the true mode shape puts additional constraints on the system, the frequency determined by Rayleigh's method is never less than the exact frequency. Generally the static deflection curve of the system gives a frequency within a few per cent of the exact value. The frequency of vibration is found by considering the conservation of energy in the system; the natural frequency is determined by dividing the expression for potential energy in the system by the expression for kinetic energy.

4.3.1 The vibration of systems with heavy springs

The mass of the spring elements can have a considerable effect on the frequency of vibration of those systems in which relatively heavy springs are used, such as in valve mechanisms and in some rotational systems.

Consider the system shown in Fig. 4.7 where a rigid body of mass M is connected to a fixed frame by a spring of mass m and stiffness k. The body moves only in the x direction.

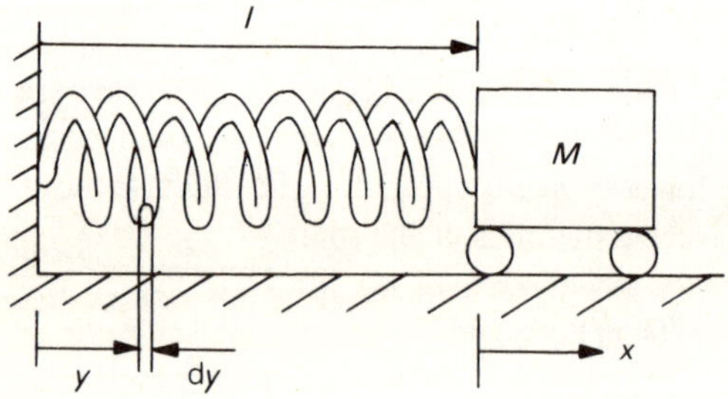

Fig. 4.7

If the dynamic deflected shape of the spring is assumed to be the same as the static shape, the velocity of the spring element is $\dot{y} = (y/l)\,\dot{x}$ and the mass of the element is $(m/l)\,\mathrm{d}y$.

Thus the kinetic energy, KE of the system is

$$\mathrm{KE} = \tfrac{1}{2}\,M\dot{x}^2 + \tfrac{1}{2}\int_0^l m/l\,((y/l)\,\dot{x})^2\,\mathrm{d}y = \tfrac{1}{2}\,(M + (m/3))\,\dot{x}^2 \quad .$$

The strain energy in the spring is $\mathrm{SE} = \frac{1}{2}\,kx^2$ and hence by assuming harmonic motion and putting $\mathrm{KE}_{\max} = \mathrm{SE}_{\max}$, (SE = PE in this case),

$$\tfrac{1}{2}\,(M + (m/3))\omega^2 X^2 = \tfrac{1}{2}\,kX^2$$

and the frequency of oscillation is $1/2\pi\,\sqrt{(k/(M + (m/3))}$ Hz, i.e. the system can be modelled by a massless spring of stiffness k with a body of mass $M + (m/3)$ attached.

4.3.2 Transverse vibration of a beam

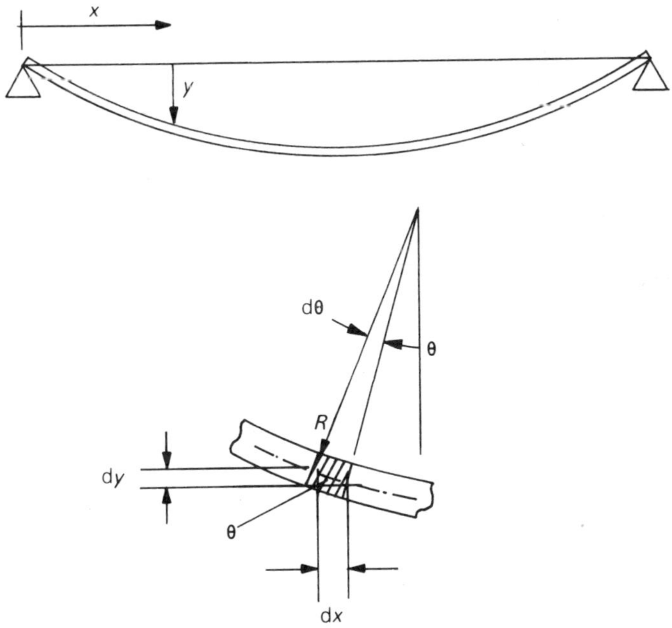

Fig. 4.8

For the beam shown in Fig. 4.8, if m is the mass/length and y is the amplitude of the assumed deflection curve

$$\text{KE}_{\text{max}} = \tfrac{1}{2}\int \dot{y}^2_{\text{max}}\ dm = \tfrac{1}{2}\,\omega^2\int y^2\ dm$$

where ω is the natural frequency of the beam corresponding to the mode assumed in the deflected shape.

The strain energy SE of the beam is the work done on the beam which is stored as elastic energy. If the bending moment is M and the slope of the elastic curve is θ,

$$\text{SE} = \tfrac{1}{2}\int M\text{d}\theta \quad .$$

Usually the deflection of beams is small so that the following relationships can be assumed to hold.

$$\theta = \text{d}y/\text{d}x \text{ and } R\ \text{d}\theta = \text{d}x,$$

i.e. $$1/R = \text{d}\theta/\text{d}x = \text{d}^2y/\text{d}x^2 \quad .$$

From beam theory $M/I = E/R$ or $1/R = M/EI$ where R is the radius of curvature and EI is the flexural rigidity.

$$\text{Thus SE}_{\text{max}} = \tfrac{1}{2}\int (M/R)\,\mathrm{d}x = \tfrac{1}{2}\int (M^2/EI)\,\mathrm{d}x$$

$$= \tfrac{1}{2}\int EI(\mathrm{d}^2y/\mathrm{d}x^2)^2\,\mathrm{d}x \quad .$$

Putting $\text{KE}_{\text{max}} = \text{SE}_{\text{max}}$ gives

$$\omega^2 = \frac{\int EI(\mathrm{d}^2y/\mathrm{d}x^2)^2\,\mathrm{d}x}{\int y^2\,\mathrm{d}m}$$

This expression gives the lowest natural frequency of transverse vibration of a beam if the static deflected shape or something similar is assumed.

Example 21

A simply supported beam of length l and mass m_2 carries a body of mass m_1 at its mid-point. Find the lowest natural frequency of transverse vibration.

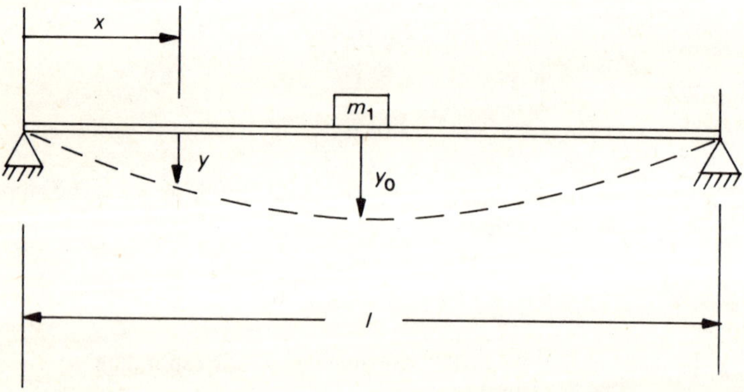

The boundary conditions are $y = 0$ and $\mathrm{d}^2y/\mathrm{d}x^2 = 0$ at $x = 0$ and $x = l$. These conditions are satisfied by a half sine wave. A polynomial expression can be derived for the deflected shape but the sinusoid is usually easier to manipulate:

$$y = y_0 \sin(\pi x/l) \quad .$$

Hence $\dot{y} = \dot{y}_0 \sin(\pi x/l)$ and $\mathrm{d}^2y/\mathrm{d}x^2 = -y_0(\pi/l)^2 \sin(\pi x/l)$

$$\int_0^l EI(\mathrm{d}^2y/\mathrm{d}x^2)^2\,\mathrm{d}x = \int_0^l EI\,y_0^2(\pi/l)^4 \sin^2(\pi x/l)\,\mathrm{d}x$$

$$= EI\,y_0^2(\pi/l)^4(l/2)$$

$$y^2\,dm = \int_0^l y_0^2 \sin^2(\pi x/l)\,(m_2/l)\,dx + y_0^2 m_1$$

$$= y_0^2\,(m_1 + (m_2/2)) \quad .$$

$$\text{Thus } \omega^2 = \frac{EI(\pi/l)^4\,(l/2)}{(m_1 + (m_2/2)} \quad .$$

$$\text{If } m_2 = 0,\ \omega^2 = (EI/2)\,(\pi^4/l^3 m_1) = 48.7\ EI/(m_1 l^3) \quad .$$

The exact solution is 48 $EI/(m_1 l^3)$, so the Rayleigh method solution is 1.4% high.

Alternatively the Dunkerley method can be used. Here $P_1^2 = 48\ EI/(m_1 l^3)$ and $P_2^2 = EI\,\pi^4/(m_2 l^3)$. Thus

$$1/\omega^2 = (m_1 l^3)/48\ EI)) + (m_2 l^3)/(\pi^4 EI) \quad .$$

$$\text{Hence } \omega^2 = \frac{EI(\pi/l)^4\ (l/2)}{(1.015\ m_1 + (m_2/2))}$$

which is very close to the value determined by the Rayleigh method.

Example 22

Find the lowest natural frequency of transverse vibration of a cantilever of mass m which has a rigid body of mass M attached at its free end.

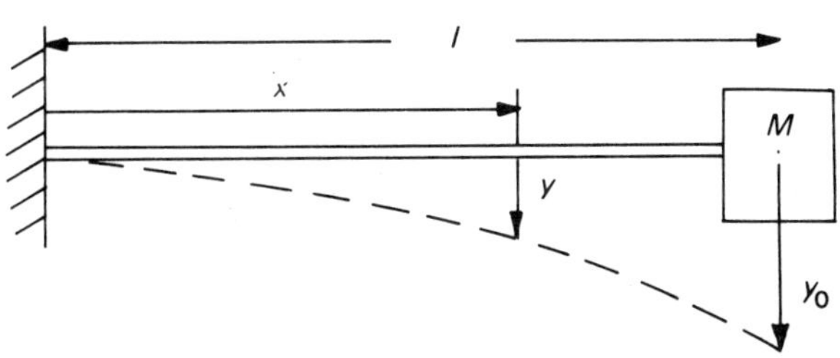

The static deflection curve is $y = (y_0/2l^3)\,(3lx^2 - x^3)$.
Alternatively $y = y_0(1 - \cos(\pi x/2l))$ could be assumed.

$$\int_0^l EI\,(d^2y/dx^2)^2\,dx = EI \int_0^l (y_0/(2l^3))^2\,(6l - 6x)^2\,dx = (3EI/l^3)\,y_0^2$$

$$\int y^2\,\mathrm{d}m = \int_0^l y^2 (m/l)\,\mathrm{d}x + y_0{}^2 M$$

$$= \int_0^l (y_0{}^2/4l^6)\,(m/l)\,(3lx^2 - x^3)^2\,\mathrm{d}x + y_0{}^2 M$$

$$= y_0^2\,\{M + (33/140)\,m\}\ .$$

Thus $\omega^2 = \{3EI/((M + (33/140)m)l^3)\}$.

4.4 THE STABILITY OF VIBRATING SYSTEMS

If a system is to vibrate about an equilibrium position, it must be stable about that position. That is, if the system is disturbed when in an equilibrium position, the elastic forces must be such that the system vibrates about the equilibrium position. Thus the expression for ω^2 must be positive, if a real value of the frequency of vibration about the equilibrium position is to exist, and hence the potential energy (PE) in a stable system must also be positive.

For conservative systems the principle of minimum PE can be used to test the stability of a system. Thus a system will be stable at an equilibrium position if the PE of the system is a minimum at that position. This requires that

$$\mathrm{d(PE)}/\mathrm{d}q = 0 \text{ and } \mathrm{d}^2\mathrm{(PE)}/\mathrm{d}q^2 > 0$$

where q is a generalised coordinate.

Hence the conditions necessary for vibration to take place are found, and the position about which vibration occurs is determined.

CHAPTER 5

Automatic control systems

Automatic control systems have been in use for several centuries, but, as with much of technology, the most rapid developments have taken place during the last fifty years. Windmills were the spawning ground of several automatic control systems, for example keeping the sails into the wind by means of a fantail (1745), the application of centrifugal governors to millstones (1783), and roller reefing (1789). In the late eighteenth century centrifugal governors were also being used to control the speed of steam engines. In the early twentieth century the rapid development of automatic control systems took place because of the need to position guns and ships quickly and accurately, which led to the development of servomechanisms in the mid 1930s. A fast rate of progress has been maintained to the present day, so that only an introduction to automatic control systems can be attempted in a text of this length.

In principle many variables can be controlled by humans, but in practice this may be impossible, difficult, costly, or undesirable because of the need for continuous operation regardless of environment, large forces, and fast response. The human reaction time of about 0.3 s is too slow for many control systems. Examples of control systems are liquid level control by ball valve, temperature control by thermostat, and surface control such as a ship's rudder or aircraft flaps by hydraulic servo. All types of control system can be modelled, irrespective of the operating mechanism.

The essential feature of an automatic control system is the existence of a **feedback loop** to give good performance. This is a closed loop system; if the measured output is not compared with the input the loop is **open**. Usually it is required to apply a specific input to a system and for some other part of the system to respond in the desired way. The error between the actual response and the ideal response is detected and fed back to the input to modify it so that the error is reduced, as shown in Fig. 5.1. The output of a device represented by a block in a block diagram cannot effect the input to that device unless a specific feedback loop is provided.

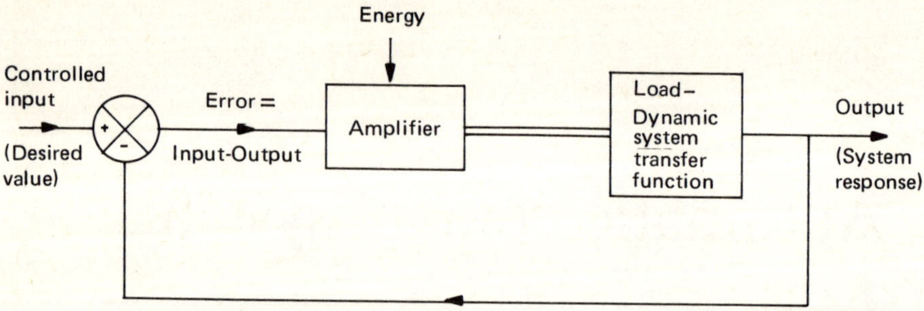

Fig. 5.1

In general both input and output vary with time, and the control system can be mechanical, pneumatic, hydraulic and electrical in operation, or any combination of these or other power sources. The system should be absolutely stable so that if excited it will settle to some steady value, and it should be accurate in the steady state.

This concept can be illustrated by considering a simple dynamic system of the type previously considered, as shown in Fig. 5.2.

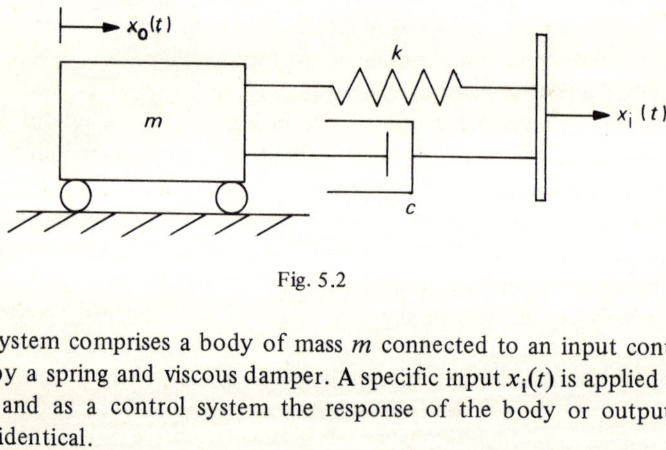

Fig. 5.2

The system comprises a body of mass m connected to an input controlled platform by a spring and viscous damper. A specific input $x_i(t)$ is applied to this platform, and as a control system the response of the body or output $x_0(t)$ should be identical.

Considering the forces acting on the body, the equation of motion is:

$$m\ddot{x}_0 = k(x_i - x_0) + c(\dot{x}_i - \dot{x}_0) \quad . \tag{5.1}$$

Equations of motion of this type have been solved for an harmonic input (equation 2.18). For a general solution irrespective of input it is convenient to use the D-operator. Thus (5.1) becomes;

$$m\,\mathrm{D}^2 x_0 = k(x_i - x_0) + c\mathrm{D}(x_i - x_0) = (k + c\mathrm{D})\,(x_i - x_0) \quad .$$

It should be noted that although using the D-operator is a neat and compact form of writing the equation it does not help with the solution of the response problem.

Now the force F on the body is $m\ D^2x_0$, so $x_0 = F/mD^2$.
This is shown in block diagram form in Fig. 5.3.

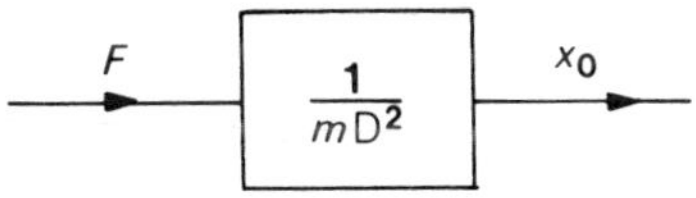

Fig. 5.3

For the spring/damper unit $F = (k + cD)\ (x_i - x_0)$ as shown in Fig. 5.4.

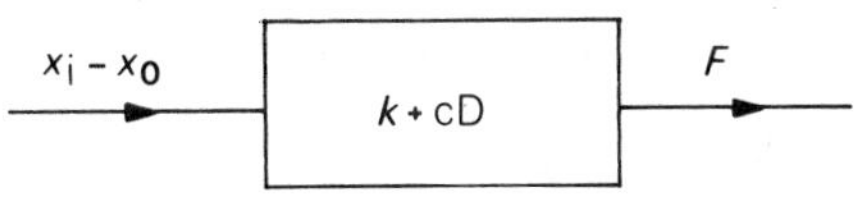

Fig. 5.4

These systems can be combined as shown in Fig. 5.5.

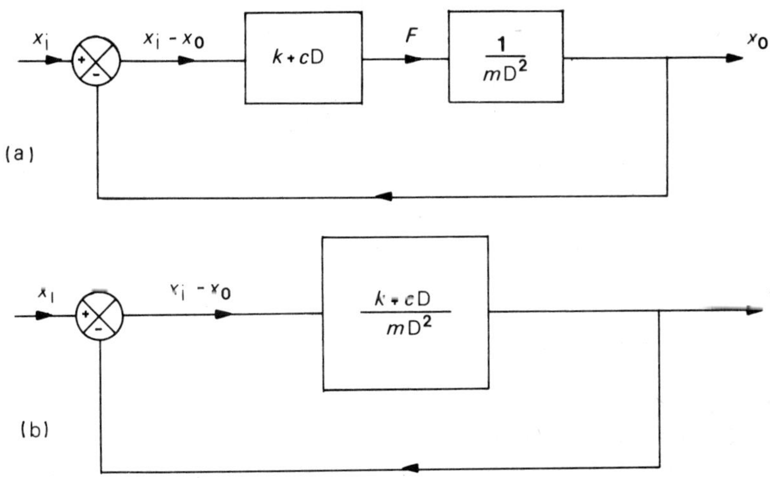

Fig. 5.5

Fig. 5.5 is the conventional unity feedback loop form. Essentially the spring acts as an error-sensing device and generates a restoring force related to that error.

$$\text{Since } (x_i - x_0)\ ((k + cD)/mD^2) = x_0,$$

the response x_0 to any input x_i can be determined.

5.1 THE SIMPLE HYDRAULIC SERVO

5.1.1 Open loop servo

Fig. 5.6 shows an open loop hydraulic servo. The input controls the position of the spool valve which directs fluid under pressure through a port to one end of the working cylinder. The working piston is connected to the controlled element so that its position may be considered to be the output.

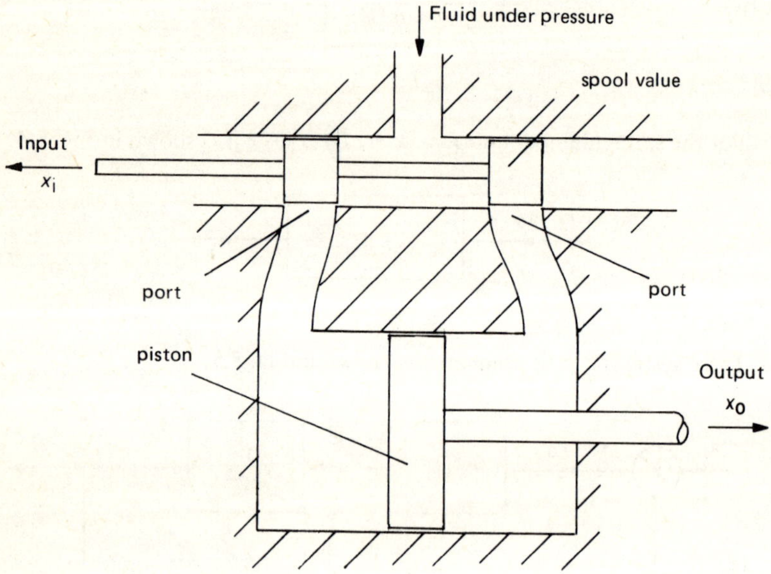

Fig. 5.6

If the area of the piston is A and the port coefficient is b, the flow equation gives

$$A \,.\, \mathrm{D}x_0 = b \,.\, x_i, \text{ i.e. } x_0 = (b/(A\ \mathrm{D}))\,x_i\ .$$

Therefore the system integrates the input.

For example, if a step input is applied, $x_i = X$ at $t \geqslant 0$, and $x_i = x_0 = 0$ when $t < 0$, thus for $t > 0, x_0 = (b/(A\ \mathrm{D}))\,x = (b/A)\,Xt$. i.e. the output increases with time.

This response is undesirable in a control system. However, some improvements can be made; for example feedback can be applied.

5.1.2 Closed loop servo

Because of the limitations of the open loop servo performance, feedback of the output is usually added as shown in Fig. 5.7. As the output reaches the desired value it acts to move the spool valve and thereby close the port.

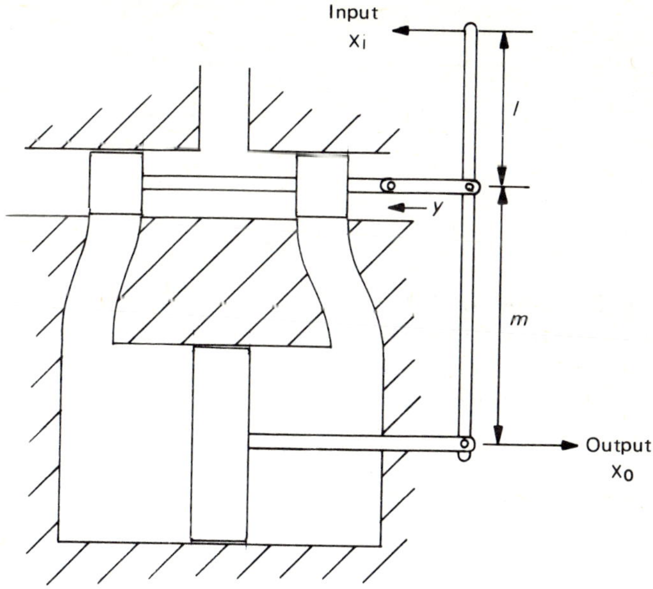

Fig. 5.7

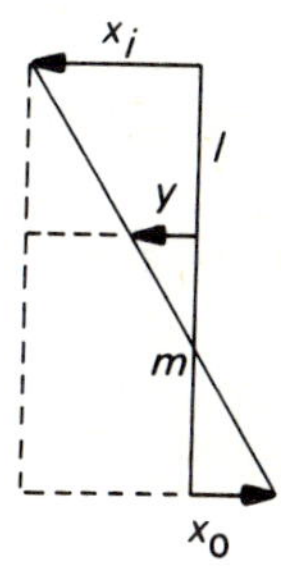

For the spool valve displacement y, $\dfrac{x_i + x_o}{l+m} = \dfrac{x_i - y}{l}$

thus, $y = x_i - (l/(l+m))\,(x_o + x_i)$

The flow equation gives $by = A\,\mathrm{D}x_o$. Thus if y is eliminated,

$$A\,\mathrm{D}x_o = -(bl/(l+m))\,x_o + (bm/(l+m))\,x_i \quad .$$

i.e. $\{1 + ((l+m)/bl)\,A\,\mathrm{D}\}\,x_o = (m/l)\,x_i$

or $\{1 + T\,\mathrm{D}\}\,x_o = (m/l)\,x_i$ (5.2)

where $T = ((l+m)/bl)\,A$ and has dimensions of time.

T is known as the **time constant** of the system. Equation (5.2) is the equation of motion and it relates the output to the input. It is sometimes convenient to consider the error ϵ between input and output where $\epsilon = x_i - x_o$.

Thus from (5.2), $(1 + T\,\mathrm{D})x_o = (m/l)\,(\epsilon + x_o)$

or $x_o = (1 - (m/l) + T\,\mathrm{D})^{-1}\,(m/l)\,\epsilon$.

Hence a block diagram can be drawn to represent the system, as shown in Fig. 5.8.

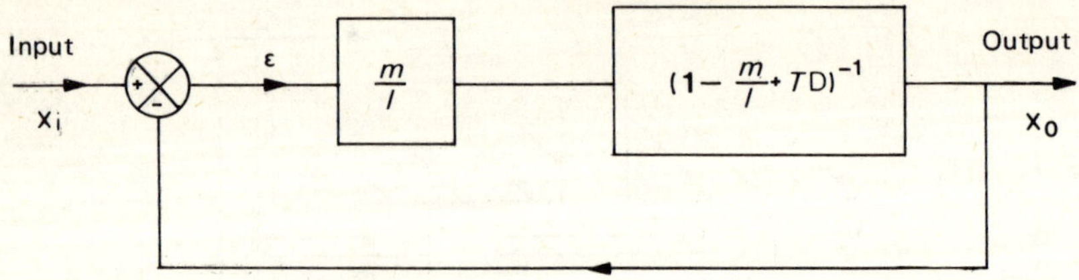

Fig. 5.8

It should be noted that in the idealised hydraulic servo considered many factors have been neglected such as friction, slackness in connections, non-linearity in the spool valve, pressure losses, compressibility of the fluid and flexibility of the metal parts. In a real system some of these factors may have to be taken into account if they have a significant effect on the system response.

Response to a Step Input

A step input is when x_i is suddenly increased to a value, say X, and when $t < 0$ $x_i = 0$ and when $t \geqslant 0$ $x_i = X$.

From equation (5.2), for the output x_0 the P.I. $= (m/l)\,X$ and C.F. $= B\mathrm{e}^{-t/T}$

Thus $x_0 = (m/l)\,X + B\mathrm{e}^{-t/T}$.

Since $x_0 = 0$ when $t = 0, B = -(m/l)\,X$, hence $x_0 = (m/l)\,X\,\{1 - \mathrm{e}^{-t/T}\}$.

i.e. x_0 increases exponentially with time to a value $(m/l)\,X$; the system time constant T has considerable effect on this response.

For the servo without feedback $x_0 = (b/A)\,X\,t$ and the output steadily increases with time.

Response to a Ramp Input

A ramp input is an input which increases uniformly with time. Thus $x_i = \beta t$ is a ramp input.

From equation (5.2), for the output x_0

P.I. $= (m/l)\,\beta t - (m/l)\,\beta$ and C.F. $= B\,\mathrm{e}^{-t/T}$

Thus $x_0 = B\,\mathrm{e}^{-t/T} + (m/l)\,\beta t - (m/l)\,\beta$.

If $x_0 = 0$ when $t = 0, B = (m/l)\,\beta$. Hence $x_0 = (m/l)\,\beta\,(\mathrm{e}^{-t/T} - 1) + (m/l)\,\beta t$.

For the case $m = l$, $x_0 = \beta(\mathrm{e}^{-t/T} - 1) + \beta t$. Thus x_0 increases exponentially with time towards a value $\beta t (= x_i)$. The error between output and input is $\beta(\mathrm{e}^{-t/T} - 1)$. Even when $t = \infty$ this error is $-\beta$. This is known as a **steady state error** because it persists after steady conditions have been attained.

Response to a Sinusoidal Input

For a sinusoidal input $x_i = X_i \cos vt$.

For x_0 the C.F. is an exponential decay, and this can usually be neglected. A P.I. can be assumed, $x_0 = X_0 \cos(vt - \phi)$.

Substituting this solution into (5.2) gives:

$$X_0 \cos(vt - \phi) - v\,T\,X_0 \sin(vt - \phi) = (m/l)\,X_i \cos vt.$$

This equation can be solved by drawing a phasor diagram as shown in Fig. 5.9.

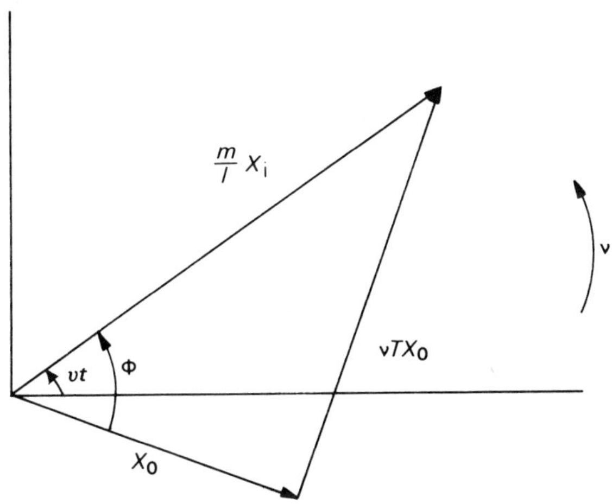

Fig. 5.9

Thus $(m/l)^2 X_i^2 = (1 + v^2T^2)\,X_0^2$

or $X_0/X_i = (m/l)/\sqrt{(1 + v^2T^2)}$ and $\phi = \tan^{-1} vT$. (5.3)

It is often desirable to show graphically how the steady state response to a sinusoidal input varies with frequency. There are several methods of doing this; one is to draw a Bode Diagram. The **Bode Diagram** is a plot of gain (X_0/X_i) against frequency. The magnitude of (X_0/X_i) is plotted on a log scale, usually dB.

The gain (or attenuation) in dB = $20 \log_{10}(X_0/X_i)$, so for equation (5.3)

$$\text{Gain (dB)} = 20 \log_{10}(m/l) - 10 \log_{10}\{1 + v^2T^2\}\ .$$

Fig. 5.10 shows the diagram for the case $m = l$.

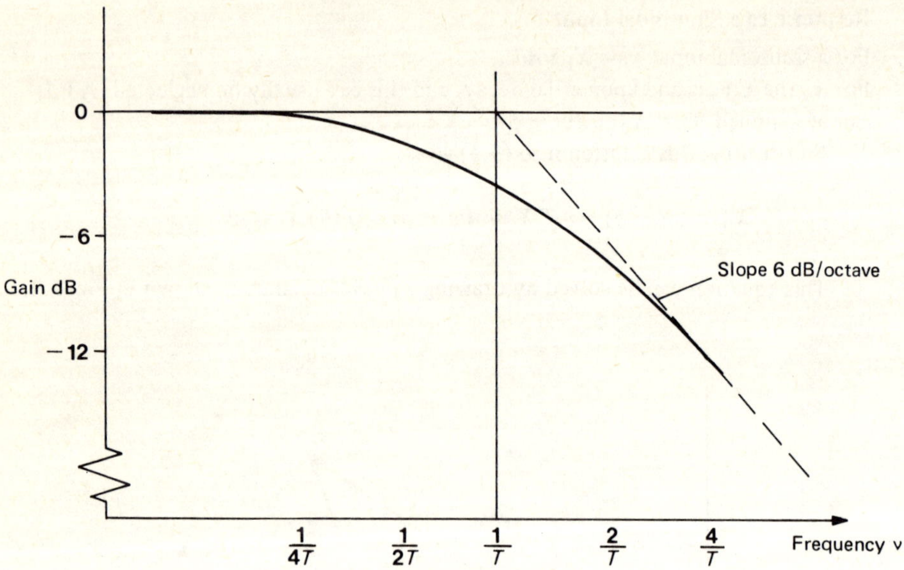

Fig. 5.10

An alternative diagram is the harmonic response locus which can be drawn by plotting gain and phase as a polar diagram.

$$\text{Since } X_i = (l/m)\, X_0\sqrt{(1 + v^2T^2)} = \{(lX_0/m)^2 + (lX_0vT/m)^2\}^{1/2}$$

the locus is as shown in Fig. 5.11.

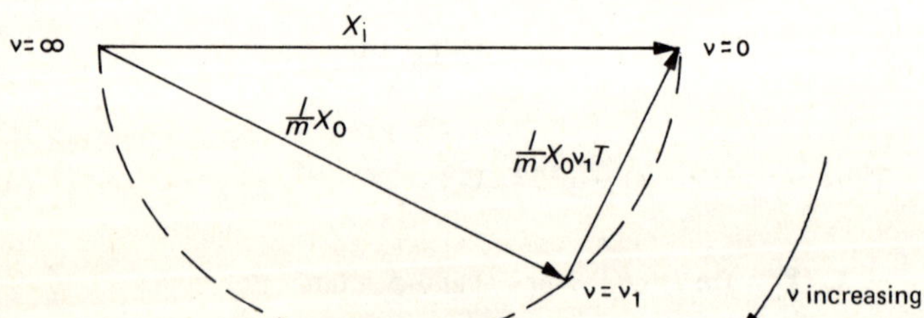

Fig. 5.11

5.2 MODIFICATIONS TO THE SIMPLE HYDRAULIC SERVO

5.2.1 Derivative control

An intelligent human controller would take account not only of the instantaneous value of the input but also its rate of change, or the rate of change of error. This would help to avoid overshooting and would improve the system response. The simple servo can be modified to act in a similar way, as shown in Fig. 5.12.

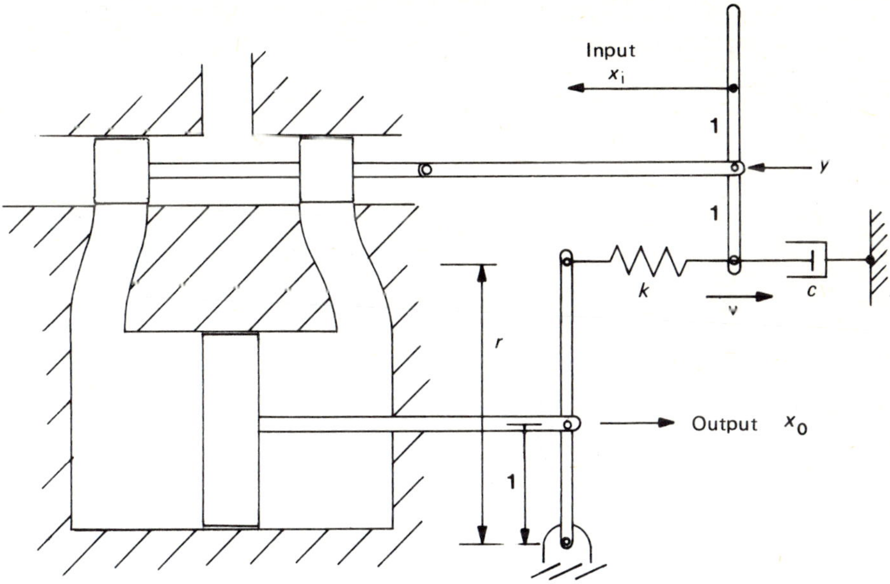

Fig. 5.12

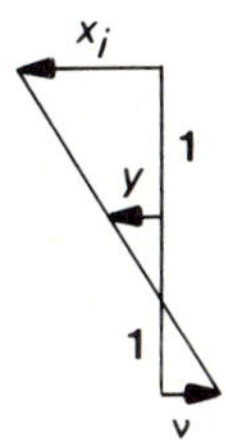

The valve displacement $y = (x_i - v)/2$. (5.4)

The flow equation gives $by = A \, \mathrm{D} \, x_0$. (5.5)

And force balance gives $k(r \, x_0 - v) - c \, \mathrm{D} \, v = 0$. (5.6)

It is required to eliminate v and y to give x_0 as $f(x_i)$.

From (5.6), $v = kr/(k + c\mathrm{D}) \, . \, x_0$.

Substitute in (5.4): $y = (x_i/2) - (kr/(k + c \, \mathrm{D})) \, (x_0/2)$

and in (5.5): $A \, \mathrm{D} x_0 = bx_i/2 - (b/2) \, (kr/(k + c \, \mathrm{D})) \, x_0$

i.e. $(2 \, A \, c \, \mathrm{D}^2 + 2 \, k \, A \, \mathrm{D} + b \, kr) \, x_0 = (b \, k + b \, c \, \mathrm{D}) \, x_i$

or $((2 \, Ac/b \, k \, r) \, \mathrm{D}^2 + (2 \, A/br) \, \mathrm{D} + 1)) \, x_0 = (1 + c/k \, \mathrm{D}) \, x_i/r$. (5.7)

Comparing (5.7) with (5.2) which was derived for a similar system without a velocity element, it can be seen that both sides of the equation of motion have gone up by one order. Thus when x_i is changed the valve setting y is not reduced by the full value corresponding to the change in x_0 because the faster x_i and x_0

change, the less y is reduced. The C.F. of x_0 is a damped oscillation or exponential decay, so the response to a step input X_i is as shown in Fig. 5.13.

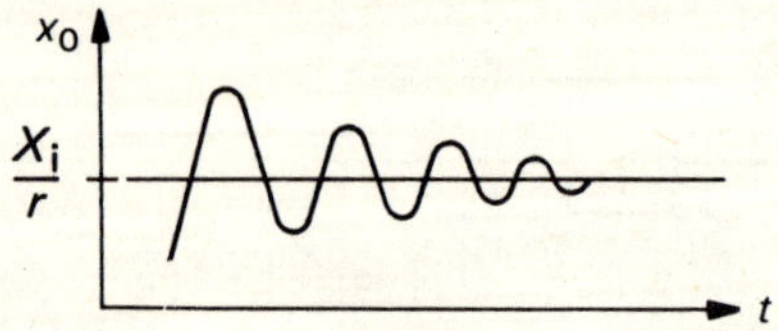

Fig. 5.13

The steady state response to a sinusoidal input can be found by drawing a phasor diagram. If $x_i = X_i \cos vt$, $x_0 = X_0 \cos(vt - \phi)$, then from Fig. 5.14.

$$X_0 = \frac{\sqrt{(1 + (cv/k)^2)}}{\sqrt{((r - (2Ac/bk)\,v^2)^2 + (2\,Av/b)^2}}\,X_i \quad .$$

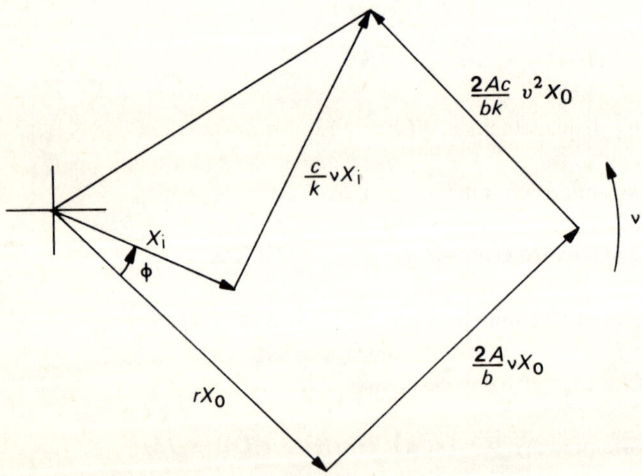

Fig. 5.14

At low frequencies $v \triangleq 0$ so $X_0 \to X_i/r$ and $\phi \to 0$.
At high frequencies $v \triangleq \infty$ so $X_0 \to 0$ and $\phi \to \pi/2$.

A detailed comparison of the simple system performance with that for a system with derivative control added depends on the system parameters, but, in general, derivative control allows a faster transient response, with overshoot if too fast.

5.2.2 Integral control

Both the simple system and that with derivative control will give steady state errors when the input is a ramp, $x_i = \beta t$. Real systems, in which fewer assumptions are made than here, can give steady state errors with other kinds of input. To eliminate these errors a correction can be added to the system which is proportional to the time integral of the error: such a system is shown in Fig. 5.15. It can be seen that compared to the system with derivate action, Fig. 5.12, k and c have been interchanged.

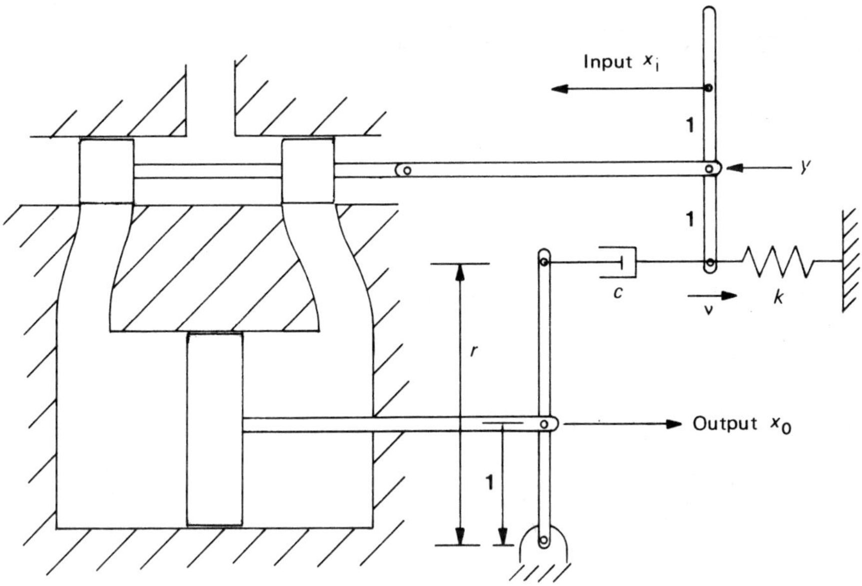

Fig. 5.15

Equations (5.4) and (5.5) apply, but force balance gives $c\mathrm{D}(rx_0 - v) - kv = 0$ and thus:

$$\left(1 + \frac{2\,Ac}{2\,Ak + bcr}\,\mathrm{D}\right)x_0 = \left(\frac{bc}{2\,Ak + bcr} + \frac{bk}{2\,Ak + bcr}\cdot\frac{1}{\mathrm{D}}\right)x_i \tag{5.8}$$

Comparing (5.8) with (5.2) shows an extra term proportional to the integral of x_i. The response to a step or sinusoidal input can be found from (5.8).

5.3 THE ELECTRIC POSITION SERVO

Electric position servos are widely used in control systems, an example is shown in Fig. 1.3.

5.3.1 The basic system

The elements of the basic position servo are shown in block diagram form in Fig. 5.16. The position of the load is compared to the desired position thereby creating an error signal. The amplifier and motor respond to the error signal and act on the load position.

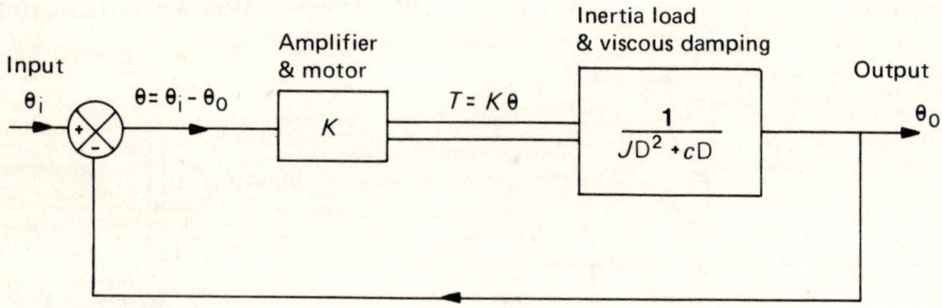

Fig. 5.16

The error signal θ is the difference between the desired position (θ_i) and the actual position (θ_o), that is, $\theta = \theta_i - \theta_o$. The amplifier and motor are such that the torque T produced by the motor is $K\theta$. For the load,

$$T = (JD^2 + c\mathrm{D})\theta_o$$

since it is considered to comprise of an inertia load with viscous damping.

$$\text{Thus} \quad K(\theta_i - \theta_o) = (J\mathrm{D}^2 + c\mathrm{D})\theta_o$$

$$\text{or} \quad (J\mathrm{D}^2 + c\mathrm{D} + K)\theta_o = K\theta_i$$

This is the same equation of motion as found for forced damped vibration.

Response of Servo to a Step Input $\theta_i = X$

For θ_o, C.F. $= A\,\mathrm{e}^{-\zeta\omega t}\sin(\omega\sqrt{(1-\zeta^2)}\,t + \phi)$

$$\text{P.I.} = \{1 + (c/K)\,\mathrm{D} + (J/K)\,\mathrm{D}^2\}^{-1}X = X$$

where $\zeta = c/c_c$ and $\omega = \sqrt{(K/J)}$ rad/s.

$$\text{Thus } \theta_o = A\mathrm{e}^{-\zeta\omega t}\sin(\omega\sqrt{(1-\zeta^2)}\,t + \phi) + X\,. \tag{5.9}$$

The response is shown in Fig. 5.17.

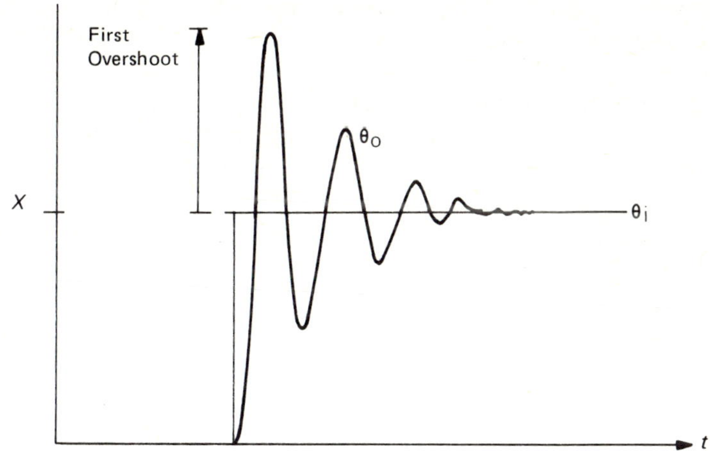

Fig. 5.17

Clearly the fastest response is when the damping is low. Unfortunately this is accompanied by a large first overshoot and a poor transient response, i.e. the oscillations take a long time to die away. Increasing the damping slows the response but reduces the overshoot and improves the transient. Critical damping gives the quickest response without overshoot.

To find the overshoot, $d\theta_o/dt$ is found and made zero and hence the times to overshoot are obtained. Substituting these times into (5.9) gives θ_o. There is no steady state error.

Response of Servo to a Ramp Input $\theta_i = \beta t$

For θ_o, C.F. as above (same system), but

$$\text{P.I.} = \{1 + (c/K)\,D + (J/K)\,D^2\}^{-1}\beta t = \beta t - (c/K)\,\beta \quad .$$

Hence there is a steady state error $(c/K)\,\beta$.

Thus high damping to achieve a good transient response also results in a large steady state error to a ramp input.

Response of Servo to an Harmonic Input $\theta_i = A\,\sin\upsilon t$

The analysis is exactly as for the forced damped vibration system (section 2.3.1)so:

$$\theta_o = KA/\sqrt{((K - J\upsilon^2)^2 + c^2\upsilon^2)}\,\sin(\upsilon t - \phi),$$

i.e. there are gain and phase errors which are functions of the system parameters and the frequency of excitation.

Because of the shortcomings of the basic system, it is often modified to provide an improved response.

5.3.2 System with output velocity feedback

If viscous damping is not enough, the damping effect can be simulated by providing a signal proportional to $D\theta_o$. This is shown in Fig. 5.18.

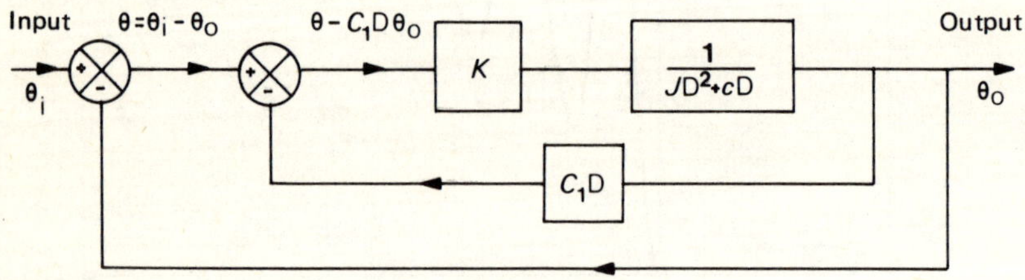

Fig. 5.18

For this system

$$K(\theta_i - \theta_o - C_1 D\theta_o) = (JD^2 + cD)\theta_o$$

i.e. $$(JD^2 + (c + KC_1)D + K)\theta_o = K\theta_i \ .$$

Thus the value of C_1 can control the total damping, and the step and transient responses can be improved at the expense of the steady state error with a ramp input.

However, if $C_1 D\theta_o$ is added to θ instead of subtracted, the opposite is true, but there is a danger of negative damping occurring with the associated unstable response.

5.3.3 System with derivative of error control

This system is shown in Fig. 5.19.

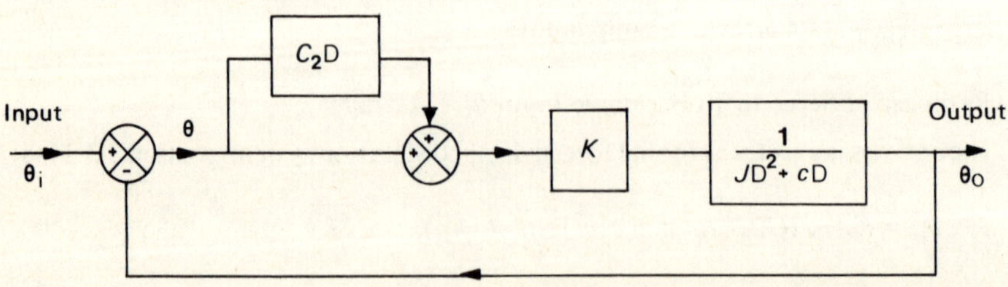

Fig. 5.19

For this system

$$K(\theta_i - \theta_o + C_2 D(\theta_i - \theta_o)) = (JD^2 + cD)\theta_o$$

i.e. $$(JD^2 + (c + KC_2)D + K)\theta_o = K(1 + C_2 D)\theta_i \quad .$$

For the response to a ramp input $\theta_i = \beta t$,

$$\text{P.I.} = \frac{K(1 + C_2 D)\beta t}{K\{1 + (c/K + C_2)D + \ldots)}$$

$$= \{1 - (c/K + C_2)\,D\ldots\ldots\}(\beta t + C_2\beta) \ ,$$

$$= \beta t - c/K\,\beta$$

i.e. $c/K\ \beta$ is the steady state error due to the damping; this has not been affected by derivative control. Thus derivative control is a method for temporarily increasing the total damping term and hence the rate of convergence of the transient response, without increasing the physical damping or the steady state error.

5.3.4 System with integral of error control

This system is shown in Fig. 5.20.

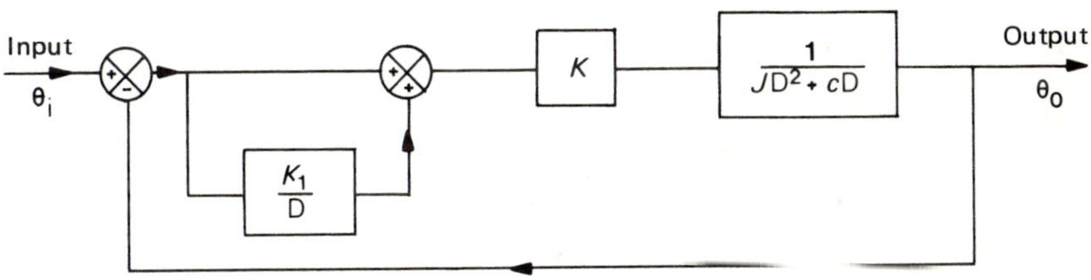

Fig. 5.20

For this system

$$K(\theta_1 - \theta_o + K_1/\text{D}\,(\theta_i - \theta_o)) = (JD^2 + cD)\theta_o$$

i.e. $$(JD^2 + cD + K + KK_1/D)\theta_o = K(1 + K_1/D)\theta_i$$

or $$(JD^3 + cD^2 + KD + KK_1)\theta_o = K(D + K_1)\theta_i \quad .$$

For the response to a ramp input $\theta_i = \beta t$,

$$\text{P.I.} = \frac{K(D + K_1)}{K(K_1 + D + \ldots.)}\ \beta t$$

$$= (1 + D/K_1 + \ldots.)^{-1}\,(K_1\beta t + \beta)/K_1 = \beta t,$$

i.e. there is no steady state error. Although this is a very desirable quality, it is possible for this system to become unstable. This can be demonstrated by considering the response to an harmonic input.

The left-hand-side of the equation of motion is 3rd order, but the response to a sinusoidal input can be found by the phasor technique.

If $\theta_i = A \cos vt$ and $\theta_o = B \cos(vt - \phi)$ in the steady state, then the equation of motion can be represented by Fig. 5.21.

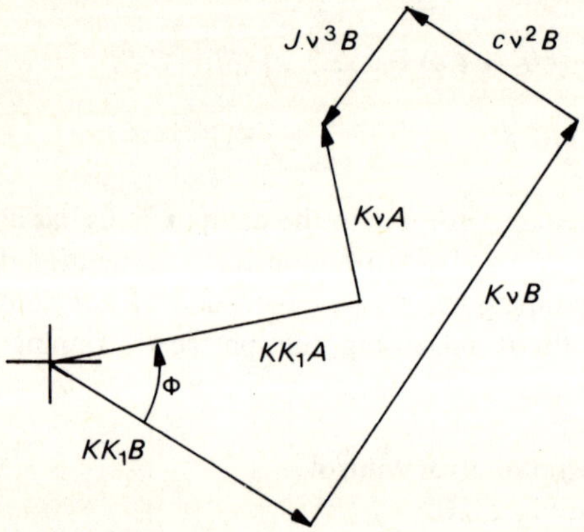

Fig. 5.21

$$\text{Thus} \quad \theta_o = \frac{A\sqrt{((Kv)^2 + (KK_1)^2)}}{\sqrt{((Kv - Jv^3)^2 + (KK_1 - cv^2)^2)}} \cos(vt - \phi) \ .$$

If the denominator is zero, B is infinite and the system is unstable. For the denominator to be zero,

$$Kv = Jv^3$$

and $KK_1 = cv^2$

Thus $c = JK_1$ and $v^2 = K/J$.

This is a critical relationship for the stable response.

The performance of the main forms of the electric position servo are compared in the table.

Table 5.1
Comparison of Main Forms of Electric Position Servo

Input		Basic system	Basic system with output velocity feedback	Basic system with derivative of error	Basic system with integral of error
Step	Response	→ Exponential decay oscillation, Rate of decay increasing			Perhaps unstable
	s.s. error	— NIL —			
Ramp	Response	→ Exponential decay oscillation, rate of decay increasing			Perhaps unstable
	s.s. error	Proportional to $D\theta_i$	Larger than $D\theta_i$	Proportional to $D\theta_i$	NIL
Harmonic	Response	Gain and phase errors proportional to system parameters and frequency of excitation.			Perhaps unstable

Example 23

In the position control servo shown in block diagram form below, the amplifier voltage gain is 100 and the error detector produces 2 V per degree of error.

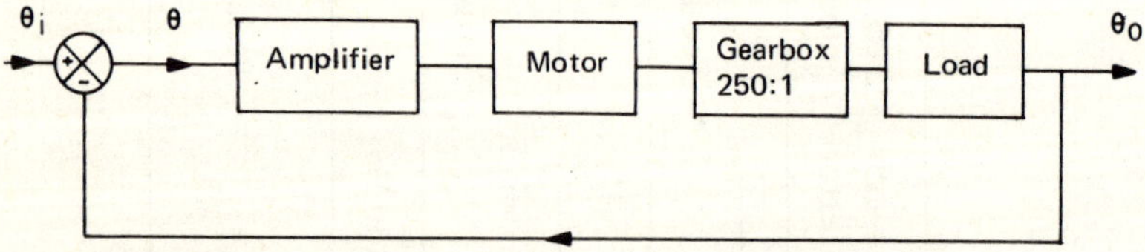

The motor produces a torque of 0.1 Nm per 100 V applied. The inertia of the load is 0.1 kg m^2 and the viscous damping is 10 Nm/rad/s. Find the damping ratio of the system.

If the input speed is 5 rad/s find the steady state error.

Motor torque = 0.1/100 × 2 × 57.3 × 100 Nm/rad error

= 11.46 Nm/rad error.

Torque applied to load = 250 × 11.46 = 2865 Nm/rad error

For damping ratio, $\zeta = c/c_c = 10/2\sqrt{(2865 \times 0.1)} = 0.295$

For $\theta_i = 5$ rad/s,

Steady state error = 10/2865 × 5 × 57.3 deg = 1 deg.

5.4 THE LAPLACE TRANSFORMATION

In the solution of differential equations of motion describing the dynamic behaviour of mechanical systems, d/dt is commonly represented by the D operator, as above. However, in general control system analysis the Laplace transformation has tended to replace other forms of differential equation representation. The use of different notations does not usually cause confusion, because the initial conditions are assumed to be zero, so the differential equation is transformed into the Laplace domain by replacing d/dt by s, d^2/dt^2 by s^2 and so on. s is the **Laplace operator** $a + jb$, where s is a complex variable, a and b are real variables, and $j = \sqrt{-1}$.

For the remainder of this chapter the Laplace transformation will be used in the analysis. For a full description of the Laplace transformation, see Engineering Science Data Unit Item No. 69025, *Solution of Ordinary Linear Differential Equations by the Laplace Transform Method.*

5.5 SYSTEM TRANSFER FUNCTIONS

The block diagram of any linear closed loop system incorporating negative feedback and having one input and one output variable can be reduced to the form shown in Fig. 5.22.

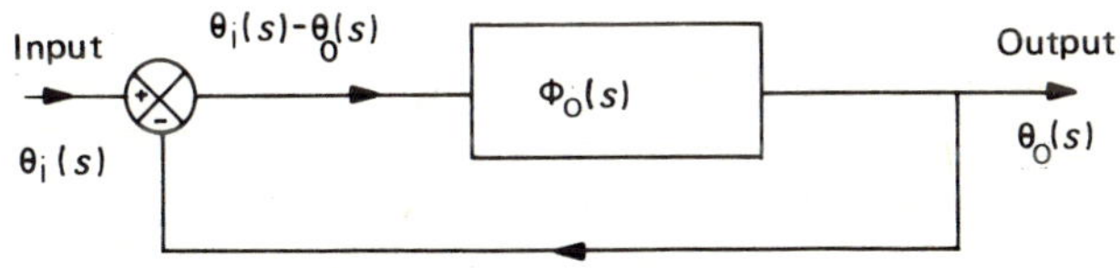

Fig. 5.22

The input and output variables have Laplace transforms $\theta_i(s)$ and $\theta_o(s)$ respectively, and the forward path of the system has a transfer function (T.F.) $\Phi_o(s)$. This is the **Open Loop Transfer Function** (O.L.T.F.) since it describes the behaviour of the system with the feedback loop open.

When the loop is closed, the input to $\Phi_o(s)$ is the error signal $\theta_i(s) - \theta_o(s)$ and thus:

$$\theta_o(s) = \Phi_o(s)\,\{\theta_i(s) - \theta_o(s)\}$$

i.e. $$\theta_o(s)/\theta_i(s) = \Phi_o(s)/1 + \Phi(s).$$

This equation determines the overall behaviour of the system when the loop is closed and $\Phi_o(s)/(1 + \Phi_o(s))$ is accordingly known as the **Closed Loop Transfer Function** (C.L.T.F.) denoted by $\Phi_c(s)$.

Example 24

Find the C.L.T.F. for the system shown below in block form.

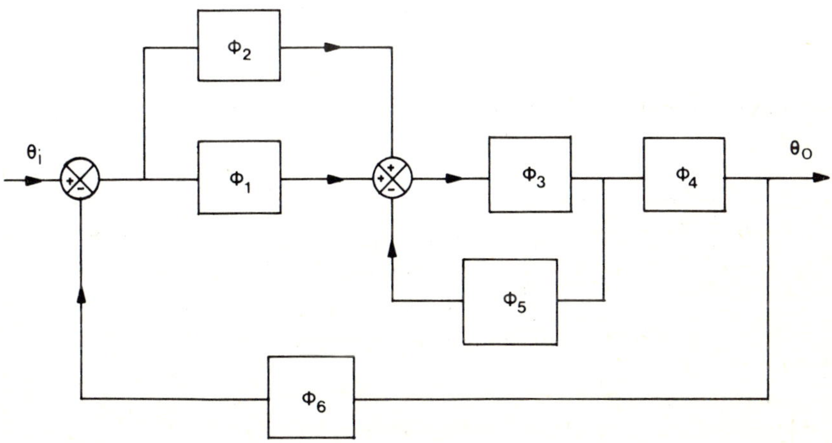

The signal leaving the first junction is:

$$\theta_i - \Phi_6 . \theta_o \quad .$$

The signal leaving the second junction is:

$$\{\theta_i - \Phi_6 . \theta_o\} (\Phi_1 + \Phi) - \Phi_5 . \theta_o / \Phi_4 \quad .$$

Thus $\{(\theta_i - \Phi_6 \theta_o)(\Phi_1 + \Phi_2) - \Phi_5 . \theta_o / \Phi_4\} \Phi_3 . \Phi_4 = \theta_o \quad .$

Hence the C.L.T.F.,

$$\frac{\theta_o}{\theta_i} = \frac{(\Phi_1 + \Phi_2)\Phi_3 . \Phi_4}{1 + \Phi_3\Phi_5 + \Phi_3\Phi_4\Phi_6(\Phi_1 + \Phi_2)}$$

For an electric position servo used for controlling the angular position of a turntable (Figs. 1.3 and 1.4) the block diagram model is as shown in Fig. 5.23.

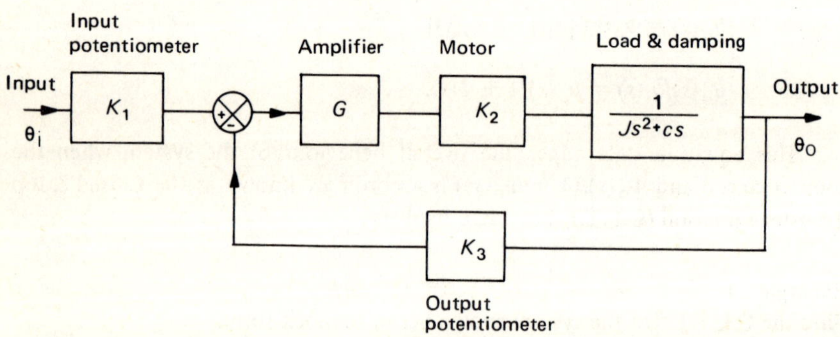

Fig. 5.23

This block diagram can be simplified as shown in Fig. 5.24.

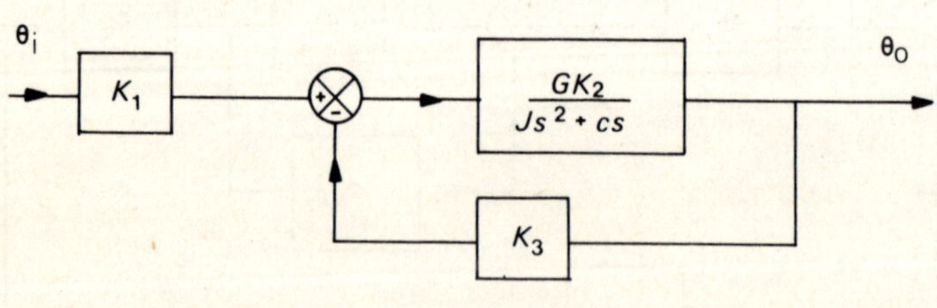

Fig. 5.24

From Fig. 5.24

$$\theta_o = (K_1\theta_i - K_3\theta_o)\,(GK_2/(Js^2 + cs)) \tag{5.10}$$

If the O.L.T.F. and C.L.T.F. are to be determined the block diagram is required in the form of Fig. 5.22.

The O.L.T.F. $\Phi_o(s) = \theta_o/(\theta_i - \theta_o)$

and C.L.T.F. $\Phi_c(s) = \theta_o/\theta_i$.

From (5.10)

$$((Js^2 + cs)/(GK_2) + K_3)\theta_o = K_1\theta_i \; .$$

Hence $\Phi_c(s) = \dfrac{GK_1K_2}{Js^2 + cs + GK_2K_3}$.

and

$$\Phi_o(s) = \frac{GK_1K_2}{Js^2 + cs + GK_2K_3 - GK_1K_2} \; .$$

It can be seen from the expression for $\Phi_c(s)$ that the frequency equation is:

$$Js^2 + cs + GK_2K_3 = 0$$

or

$$s^2 + cs/J + GK_2K_3/J = 0$$

since the values of s which satisfy this equation make $\Phi_c(s) = \infty$. These values can be denoted by p_1 and p_2 where:

$$(s - p_1)\,(s - p_2) = 0$$

$$p_1 = a + jb \text{ and } p_2 = a - jb \; .$$

Now $s = -c/2J \pm j\sqrt{(GK_2K_3/J - (c/2J)^2)}$

so $a = -c/2J$ and $b = \sqrt{(GK_2K_3/J - (c/2J)^2)}$.

These roots can be plotted on the s-plane as shown in Fig. 5.25, as G increases from zero.

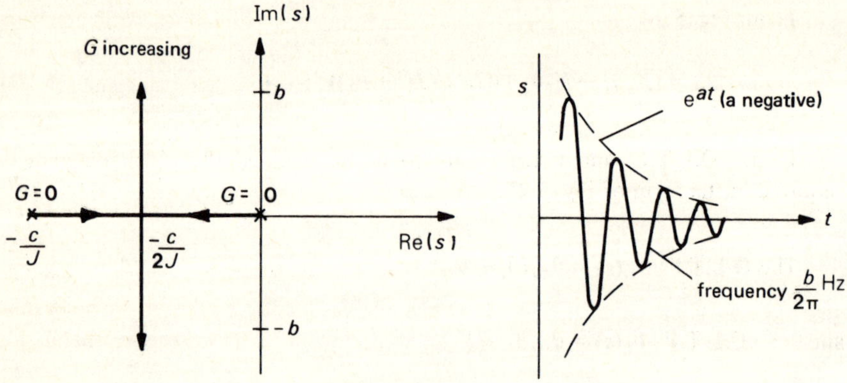

Fig. 5.25

The frequency equation of a system governs its response to a stimulus. If the roots lie on the left-hand side of the s-plane the response dies away with time and the system is stable. If the roots lie on the right-hand side of the s-plane the response grows with time and the system is unstable.

In general any transfer function $\Phi(s)$ has the form:

$$\Phi(s) = \frac{K(s^m + \alpha_1 s^{m-1} + \alpha_2 s^{m-2} + \ldots \alpha_m)}{(s^n + \beta_1 s^{n-1} + \beta_2 s^{n-2} + \ldots \beta_n)} \quad \text{and } n \geqslant m$$

with a set of roots $s_1, s_2, s_3 \ldots . s_n$.

$$\text{Thus } \Phi(s) = \frac{K(s - z_1)(s - z_2) \ldots . .(s - z_m)}{(s - p_1)(s - p_2) \ldots . .(s - p_n)}$$

where the values of $s = p_1, p_2 \ldots . . p_n$ are those which make $\Phi(s) = \infty$ and are called **poles**, and the values of $s = z_1, z_2 \ldots z_m$ are those which make $\Phi(s) = 0$ and are called **zeros**.

Hence the poles of $\Phi_c(s)$ are the natural frequencies of the system.

5.6 ROOT LOCUS

In most systems instability is intolerable. In dynamic systems a degree of damping is usually desired for safety, while in control systems a considerable margin of stability is essential, preferably with little or no oscillation. However, excessive damping is wasteful of energy.

Consider the diagram of the s-plane Figs. 3.11, 5.25; the right-hand side is a completely unstable region, the imaginary axis is neutral equilibrium, and the left-hand side is stable. In both dynamic and control systems therefore it is important to know where the roots of the frequency or characteristic equation lie on this plane, and if necessary how to adjust the system to move them.

The idea of the Root Locus technique is to use a diagram such as the s-plane to see how the roots of the frequency equation vary as various system parameters are changed. Consider the system shown in Fig. 5.22.

$$\Phi_c(s) = \theta_o/\theta_i = \Phi_o(s)/(1 + \Phi_o(s)) \; .$$

Instead of writing $\Phi_c(s) = \infty$ for the condition of resonance it is more convenient to put $1 + \Phi_o(s) = 0$: this is the frequency equation. For a system to be stable all roots of the frequency equation must lie on the left-hand side of the Imaginary Axis when plotted on the s-plane.

If $\Phi_o(s) = K\Phi_o{}'(s)$ where:

$$\Phi_o{}'(s) = \frac{(s - z_1)(s - z_2) \ldots\ldots (s - z_m)}{(s - p_1)(s - p_2) \ldots\ldots (s - p_n)}$$

the Root Locus can be sketched as K increases from zero to infinity and the conditions for instability determined. K is the overall system gain constant.

Techniques for the construction of a Root Locus for a known control system have been developed and can be summarised as a set of rules.

5.6.1 Rules for constructing root loci

Rule 1 Number of Loci
The number of root loci is equal to the degree of the characteristic equation.

Proof The number of characteristic roots is equal to the degree of the characteristic equation, and each root has its own locus.

Rule 2 Symmetry of Loci
The root loci for a real characteristic equation are symmetrical with respect to the real axis.

Proof The complex roots of a real characteristic equation occur in conjugate pairs.

Rule 3 Poles of $\Phi_o{}'(s)$
The poles of $\Phi_o{}'(s)$ lie on the root loci and correspond to $K = 0$.

Proof Since $1 + \Phi_o(s) = 0$ and $\Phi_o(s) = K\Phi_o{}'(s)$ then $\Phi_o{}'(s) = -1/K$.

Thus if $K = 0$, $\Phi_o{}'(s) = \infty$ and $(s - p_1)(s - p_2) \ldots\ldots = 0$ i.e. when $K = 0$ the roots of the characteristic equation are the poles of $\Phi_o{}'(s)$.

Rule 4 Zeros of $\Phi_o{}'(s)$
The zeros of $\Phi_o{}'(s)$ lie on the root loci and correspond to $K = \pm\infty$.

Proof Since $1/K + \Phi_o{}'(s) = 0$ as $K \to \pm\infty$, $\Phi_o{}'(s) \to 0$ i.e. $(s - z_1)(s - z_2) \ldots\ldots = 0$.

Rule 5 Asymptotes of Root Loci
If $\Phi_o{}'(s)$ has r more poles than zeros, the root loci are asymptotic to r straight lines making angles $(2N+1)\pi/r$ with the real axis ($N = 0, 1, 2 \ldots\ldots r-1$), and also to r straight lines making angles $2N\,\pi/r$ with the real axis. The root loci approach the former asymptotes when $k \to +\infty$ and the latter when $k \to -\infty$.

Proof

$$\Phi_o{}'(s) = \frac{s^m + \alpha_1 s^{m-1} + \ldots\ldots \alpha_m}{s^n + \beta_1 s^{n-1} + \ldots\ldots \beta_n} \quad .$$

Since $\quad 1 + K\Phi_o{}'(s) = 0, \quad \text{if } r = n - m > 0$

$$1/\Phi_o{}'(s) = s^r + (\beta_1 - \alpha_1)s^{r-1} \;\ldots\ldots = -K$$

i.e. $|K| \to \infty$ when $|s| \to \infty$. Furthermore, since the first two terms dominate the expression for $1/\Phi_o{}'(s)$ when $|s| \to \infty$, this equation is approximately:

$$(s - a_o)^r = -K \begin{matrix} = |K| e^{j(2N+1)\pi} & \text{for } K > 0 \\ = |K| e^{j2n\pi} & \text{for } K < 0 \end{matrix}$$

where $a_o = -(\beta_1 - \alpha_1)/r$ and N is any integer.

Each of these equations has r distinct solutions given by:

$$s - a_o = \begin{matrix} |K|^{1/r}\, e^{j(2N+1)\pi/r} & K > 0 \\ |K|^{1/r}\, e^{j2N\pi/r} & K < 0 \end{matrix}$$

where $N = 0, 1, 2 \ldots r - 1$. Putting $s = a + jb$ and equating real and imaginary parts yields:

$$\left.\begin{aligned} a - a_o &= |K|^{1/r} \cos(2N+1)\pi/r \\ b &= |K|^{1/r} \sin(2N+1)\pi/r \end{aligned}\right\} K > 0$$

and

$$\left.\begin{aligned} a - a_o &= |K|^{1/r} \cos 2N\pi/r \\ b &= |K|^{1/r} \sin 2N\pi/r \end{aligned}\right\} K < 0$$

By division

$$b = \begin{matrix} (a - a_o)\tan(2N+1)\pi/r & & K > 0 \\ (a - a_o)\tan 2N\pi/r & & K < 0 \end{matrix} \; .$$

These are the equations of the asymptotes of the root loci. Each of these equations represents a family of r straight lines in the s-plane. The angular inclinations of these lines with the real axis are $(2N + 1)\pi/r$ and $2N\ \pi/r$ when $K > 0$ and $K < 0$ respectively.

Rule 6 Point of Intersection of Asymptotes
Both sets of asymptotes intersect on the real axis at a point with abscissa:

$$a_o = \left(\sum_1^n p_i - \sum_1^m z_i \right) / r$$

where p_i and z_i are the poles are zeros respectively of $\Phi_o'(s)$.

Proof From Rule 5 proof, in both cases when $b = 0, a = a_o = -(\beta_1 - \alpha_1)/r$ for all values of K. All the asymptotes therefore intersect on the real axis at the point $(a_o, 0)$. The abscissa a_o has the value quoted above since it can be deduced from the expression for $\Phi_o'(s)$ given that $\alpha_1 = -\sum_1^m z_i$ and $\beta_1 = -\sum_1^n p_i$ (Algebraic rule of roots of equations).

Rule 7 Root loci on the Real Axis
If $\Phi'(s)$ has at least one real pole or zero, the whole of the real axis is occupied by root-loci: a segment of the real axis corresponds to positive or negative values of K according to whether an odd or even number of poles and zeros of $\Phi_o'(s)$ lie to its right.

Proof The arguments of the complex numbers represented by the vectors drawn from the conjugate complex poles or zeros of $\Phi_o'(s)$ to a point on the real axis cancel out in pairs.

The argument of a vector drawn *from* a real pole or zero lying to the left of a point on the real axis is zero, whilst the corresponding quantity for a real pole or zero lying to the right is π.

$$\text{Now } \Phi_o'(s) = \frac{(s - z_1)(s - z_2)\ldots(s - z_m)}{(s - p_1)(s - p_2)\ldots(s - p_n)}$$

and $K\Phi_o'(s) = -1\ e^{j(2N+1)\pi}$ for $K > 0$

so $\arg K\Phi_o'(s) = (2N+1)\pi$ for $K>0$.

$$\text{Thus } \sum_1^m \arg(s-z_i) - \sum_1^n \arg(s-p_i) = \begin{matrix}(2N+1)\pi \text{ for } K>0 \\ 2N\pi \text{ for } K<0\end{matrix} \quad .$$

It follows that the first of these equations will be satisfied on the real axis only at those points having an odd number of poles and zeros to their right, and that the second equation will be satisfied only when this number is even. Zero is regarded as an even number.

Rule 8 Breakaway Points

Breakaway points indicate the existence of multiple characteristic roots and occur at those values of s which satisfy $dK/ds = 0$.

Consider a characteristic equation which has a root of multiplicity $q(\geqslant 2)$ at $s = s_o$ when $K = K_o$. In the root locus for such an equation q loci will converge on the point $s = s_o$ as K increases towards K_o, and will then breakaway from this common point as K increases beyond K_o. Typical breakaway points are shown in Fig. 5.26.

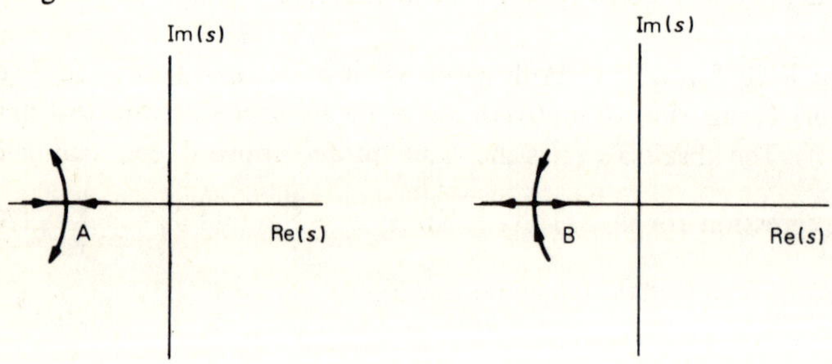

A and B each represent two equal real roots,

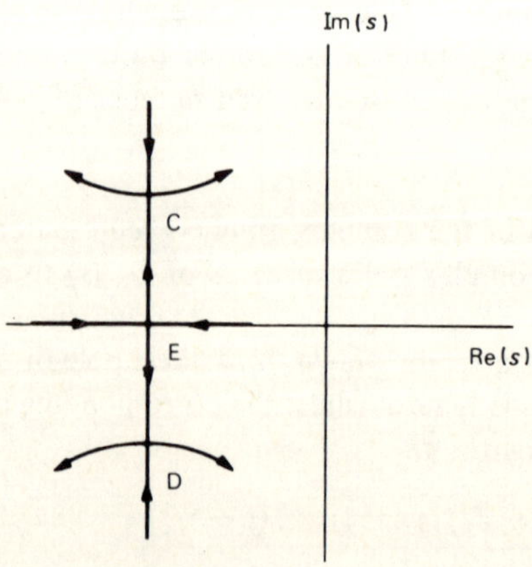

C and D each represent two equal complex roots, and E represents two equal real roots.

Fig. 5.26

Proof The characteristic equation can be written:

$$f(s, K) = P(s) + K\,Q(s) = 0 \quad .$$

Thus $\mathrm{d}f/\mathrm{d}s = \partial f/\partial s + \partial f/\partial K \,.\, \mathrm{d}K/\mathrm{d}s = 0$

and consequently $\dfrac{\mathrm{d}K}{\mathrm{d}s} = -\dfrac{\partial f/\partial s}{Q(s)} \quad .$

Clearly $\partial f/\partial K = Q(s)$

and since $f(s, K_o) = (s - s_o)^q\, g(s)$

at a breakaway point $s = s_o$, where $g(s)$ is some function such that $g(s_o) \neq 0$ it is evident that:

$$\partial f(s, K_o)/\partial s = q(s - s_o)^{q-1} g(s) + (s - s_o)^q\, g'(s)$$
$$= 0 \text{ at } s = s_o \quad .$$

Hence $(\mathrm{d}K/\mathrm{d}s)_{s=s_o} = 0$ if $Q(s_o) \neq 0 \quad .$

Rule 9 Intersections of Root Loci with the Imaginary Axis
The intersections of root loci with the imaginary axis can be found by calculating the values of K which result in existence of imaginary characteristic roots. These values of K together with the corresponding imaginary roots can be found by writing $s = j\omega$ in the characteristic equation and equating the real and imaginary parts.

For example if $F(s) = s^3 + 2s^2 + 3s + K + 2 = 0$

Putting $s = j\omega$ gives $-j\omega^3 - 2\omega^2 + 3j\omega + K + 2 = 0.$

Equate real parts: $-2\omega^2 + K + 2 = 0.$

Equate imaginary parts: $-\omega^3 + 3\omega = 0.$

Hence $\omega^2 = 3$ and $K = 4$ at intersection of loci with the imaginary axis.

Rule 10 Slopes of Root-Loci at the Complex Poles and Zeros of $\Phi_o{}'(s)$
The slope of a root locus at a complex pole or zero of $\Phi_o{}'(s)$ can be found by applying the equations

$$\sum_1^m \arg(s - z_i) - \sum_1^n \arg(s - p_i) = \begin{matrix} (2N+1)\pi \text{ for } K > 0 \\ 2N\pi \text{ for } K < 0 \end{matrix}$$

to a point in the neighbourhood of the pole or zero.

For example, consider the complex pole P_1 shown in Fig. 5.27 where γ is the unknown slope of the locus at P_1.

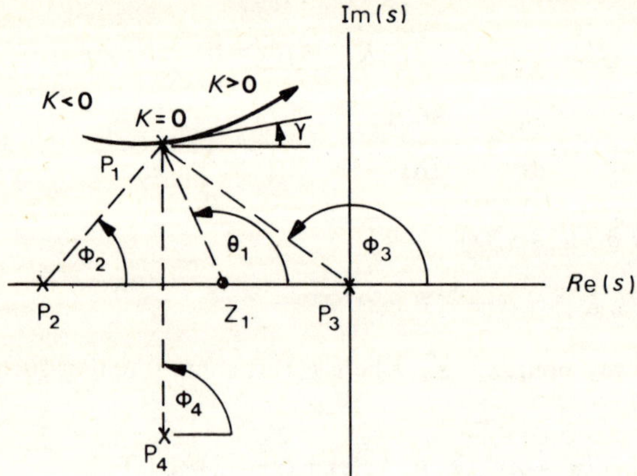

Fig. 5.27

The arguments of the complex numbers represented by the vectors drawn from the other poles P_2, P_3 and P_4 and the zero Z_1 to a point on the root locus *near* P_1 can be considered to be the angles ϕ_2, ϕ_3, ϕ_4 and θ_1.

If the first of the above equations is applied to a point in the neighbourhood of P_1 for which $K > 0$,

$$\theta_1 - \phi_2 - \phi_3 - \phi_4 - \gamma = (2N + 1)\pi$$

where n is an integer. Hence γ.

The second equation may be applied to a point near P_1 for which $K < 0$.

Rule 11 Calculation of K on the Root Loci
The absolute magnitude of the value of K corresponding to any point s_r on a root locus can be found by measuring the lengths of the vectors drawn to s_r from the poles and zeros of $\Phi_o'(s)$ and evaluating

$$|K| = \frac{(s_r - p_1)(s_r - p_2)\ldots.}{(s_r - z_1)(s_r - z_2)\ldots.}$$

The rules for constructing root loci can be summarised as follows:

1. Number of loci.
2. Symmetry of loci.
3. Poles.
4. Zeros.
5. Asymptotes.

6. Intersection of asymptotes.
7. Root loci on real axis.
8. Breakaway points.
9. Intersection with Imaginary Axis.
10. Slopes of loci at complex poles and zeros.
11. Calculation of K.

Poles are plotted x, zeros as o. It is not usually necessary to use all rules for a root locus sketch.

Example 25

Sketch the root locus for a closed-loop control system with open loop transfer function given by:

$$\frac{K}{s(s+1)(s^2+s+1)}$$

as K increases from $-\infty$ to $+\infty$. Hence determine the range of values K can have and the system be stable.

The rules are applied as follows:

1. 4 loci
2. Loci are symmetrical about the real axis for real K.
3. Poles given by $s(s+1)(s^2+s+1)=0$
   i.e. $s=0, -1, -\frac{1}{2} \pm j\sqrt{(3/2)}$. Root loci pass through poles when $K=0$.
4. No zeros.
5. $r=4$. Thus inclination of asymptotes are

   $$\frac{\pi}{4}, \frac{3\pi}{4}, \frac{5\pi}{4}, \frac{7\pi}{4} \quad \text{for } K \to +\infty$$

   $$0, \frac{2\pi}{4}, \frac{4\pi}{4}, \frac{6\pi}{4} \quad \text{for } K \to -\infty$$

6. Asymptotes intersect at

   $$\frac{0-1-\frac{1}{2}+j\sqrt{(3/2)}-\frac{1}{2}-j\sqrt{(3/2)}-0}{4} = -\frac{1}{2}$$

7. Whole of real axis occupied.

8. $\mathrm{d}K/\mathrm{d}s = 0$ gives $4s^3+6s^2+4s+1=0$.

   Breakaway points are $-\frac{1}{2}, -\frac{1}{2} \pm j\frac{1}{2}$

   When $s=-\frac{1}{2}$, $K=\frac{3}{16}$ and when $s=-\frac{1}{2} \pm j\frac{1}{2}$, $K=\frac{1}{4}$.

9. Since $1 + \Phi_o(s) = 0$,

   $s(s+1)(s^2+s+1) + K = 0$. Thus

   in $s^4 + 2s^3 + 2s^2 + s + K = 0$ put $s = j\omega$

   $\omega^4 - 2j\omega^3 - 2\omega^2 + j\omega + K = 0$

   (Real part) $\qquad \omega^4 - 2\omega^2 + K = 0$

   (Imaginary part) $-2\omega^3 + \omega = 0$,

   i.e. $\omega = 0$ and $K = 0$

   or $\omega^2 = \frac{1}{2}$ and $K = \frac{3}{4}$.

This is sufficient information to sketch the root locus, as shown in the diagram.

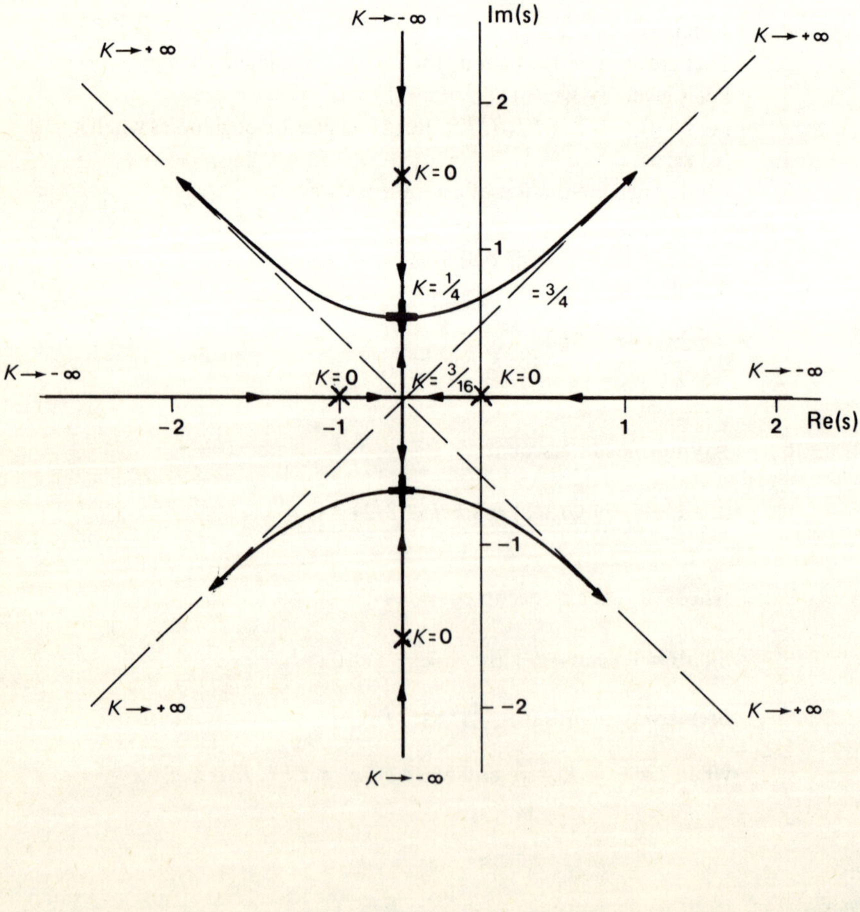

As K increases from $-\infty$ to zero the four roots move along the $0, \pi/2, \pi$ and $3\pi/2$ asymptotes to the poles ($K = 0$). As K increases from zero the loci on the real axis meet at a breakaway point ($K = 3/16$) and move to meet the other two loci at the other breakaway points ($K = 1/4$). As K increases further the four loci move off towards the $\pi/4$, $3\pi/4$, $5\pi/4$ and $7\pi/4$ asymptotes. The loci cross the imaginary axis when $K = 3/4$. Therefore the range of values K can have for the system to be stable is > 0 and $< 3/4$.

It can be seen that as K increases from 1/4 to 3/4, the frequency of oscillation of the response increases slightly and the rate of decay of oscillation decreases.

Root loci are often only drawn for positive values of the system gain K.

Example 26

Sketch the root loci of the equation

$$s^4 + 5s^3 + 8s^2 + (6 + K)s + 2K = 0$$

if K can have positive real values.

Rewrite as $1 + \dfrac{K(s+2)}{s(s+3)(s^2+2s+2)} = 0$,

so $\Phi_0{}'(s) = \dfrac{s+2}{s(s+3)(s^2+2s+2)}$.

1. Four loci.
2. Loci are symmetrical about the real axis for real K.
3. Poles given by $s(s + 3)(s^2 + 2s + 2) = 0$.

   i.e. $s = 0, -3, -1 \pm j$. Root loci pass through poles when $K = 0$.
4. Zeros given by $s + 2 = 0$ i.e. $s = -2$.

   Root loci pass through zeros when $K = \infty$.
5. $r = 3$. Thus inclination of asymptotes are $\pi/3, \pi, 5\pi/3$ for $K \to \infty$.
6. Asymptotes intersect at

$$\frac{(0 - 3 - 1 + j - 1 - j) - (-2)}{3} = -1.$$

7. Real axis occupied as below:

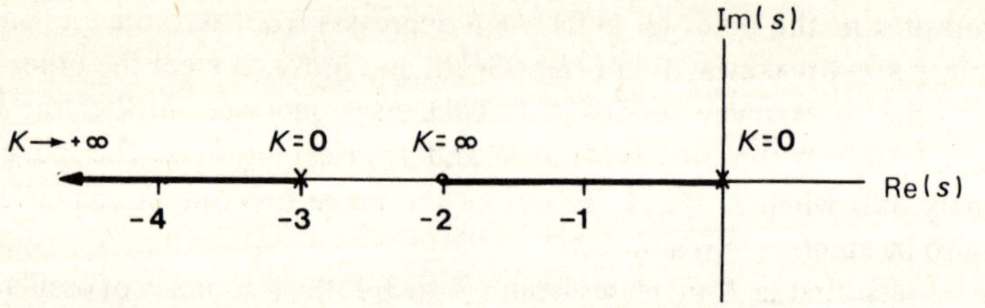

8. No breakaway points because two loci on real axis and only two other loci which are symmetrical with respect to the real axis.
9. In $s^4 + 5s^3 + 8s^2 + (6 + K)s + 2K = 0$ put $s = j\omega$.
Hence $K = 5\sqrt{13} - 11$.
10. The slope, γ, of the tangent to locus at the pole $-1 + j$ is given by:

$$45 - 30 - 90 - 135 - \gamma = (2N + 1)\,180.$$

Hence $\gamma = -30°$.

The root locus is as shown in the diagram.

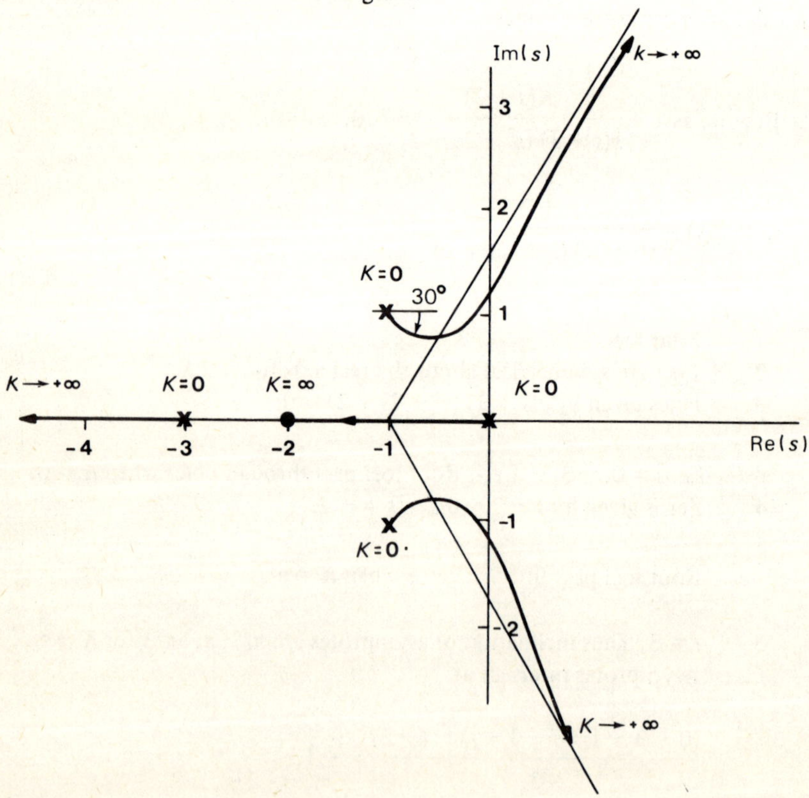

5.7 CONTROL SYSTEM FREQUENCY RESPONSE

The conditions for limiting stability were found by the root locus method by plotting $1 + \Phi_0(s) = 0$ and finding the value of K for $s = j\omega$. An alternative method is to plot $\Phi_0(s)$ with $s = j\omega$, i.e. sinusoidal forcing, and find when this quantity is equal to -1. Thus the closed loop stability is determined from the open loop response. This is the basis of the Nyquist criterion. The Nyquist criterion is an important method for studying linear feedback systems since it is expressed only in terms of the open-loop transfer function.

The open loop transfer function can be plotted on a Nyquist diagram as shown in Fig. 5.28. If the loop formed by the open loop transfer function as the frequency increases from $-\infty$ to $+\infty$ *encloses* the point $\Phi_0(s) = (-1, 0)$ the system is unstable. The proximity of the loop to that point is a measure of how stable the system is.

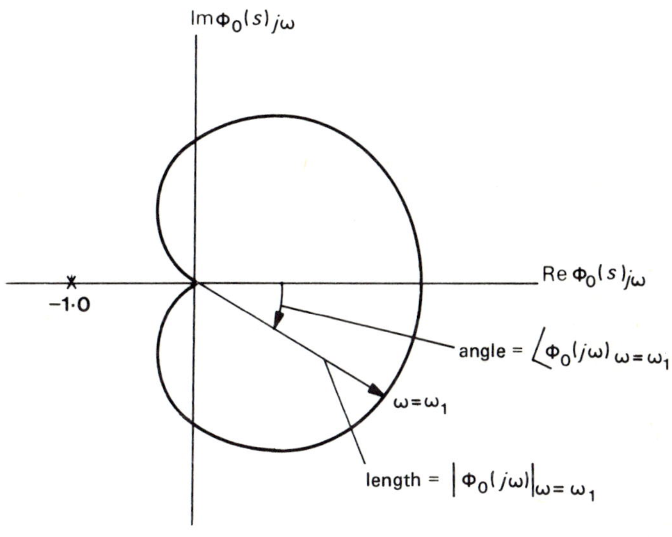

Fig. 5.28

All points to the right of a contour, as it is traversed in a clockwise direction, are said to be **enclosed** by it.

The closeness of a contour to the $(-1, 0)$ point can be expressed in terms of the **Gain Margin** and the **Phase Margin**, Fig. 5.29.

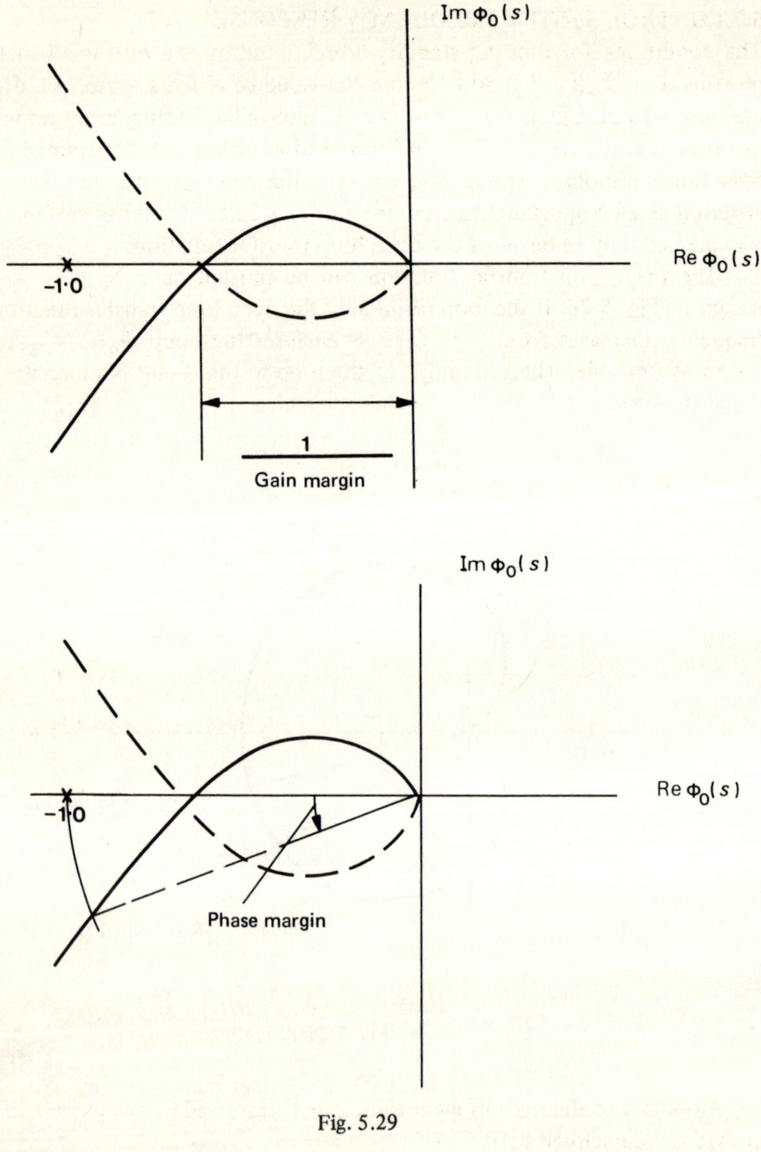

Fig. 5.29

Example 27

Draw the Nyquist Diagram for the system with an open loop transfer function

$$\frac{K}{(s+a)(s+b)}$$

and hence determine whether the closed loop system will be stable.

$$\Phi_o(s) = \frac{K}{(s+a)(s+b)}$$

$$\Phi_o(j\omega) = \frac{K}{(a+j\omega)(b+j\omega)} = \frac{K}{(ab - \omega^2) + j\omega(a+b)} .$$

Rationalise and hence

$$\Phi_o(j\omega) = \frac{K\{(ab-\omega^2) - j\omega(a+b)\}}{\{(ab-\omega^2)^2 + \omega^2(a+b)^2\}} .$$

$$\text{Thus Re } \Phi_o(j\omega) = \frac{K(ab-\omega^2)}{\{(ab-\omega^2)^2 + \omega^2(a+b)^2\}}$$

$$\text{and Im } \Phi_o(j\omega) = \frac{-K\,\omega(a+b)}{\{(ab-\omega^2)^2 + \omega^2(a+b)^2\}} .$$

Hence the following table can be deduced

ω	Re $\Phi_o(j\omega)$	Im $\Phi_o(j\omega)$
$-\infty$	0	0
$-$ve large	$-$ ve	+ ve
$-$ve small	+ ve	+ ve
0^-	+ ve	0

From this table the Nyquist diagram can be sketched, as shown.

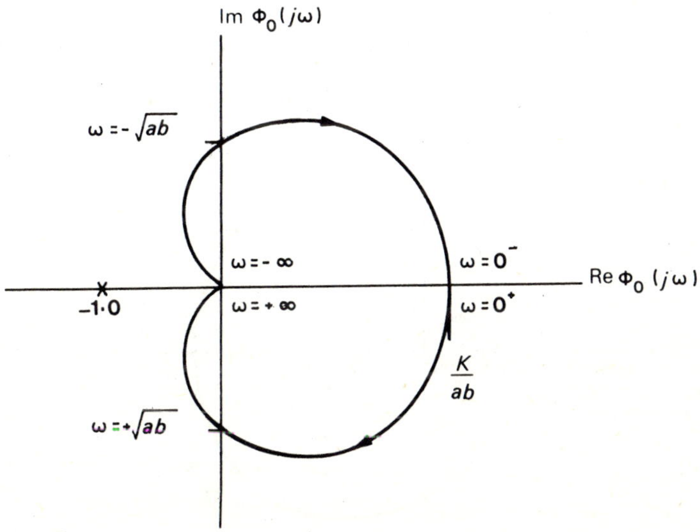

It can be seen that increasing K merely increases the size of the loop, and it will never pass through $(-1, 0)$. Thus the system is always stable.

Example 28

Consider the closed loop system with an open loop transfer function given by:

$$\frac{K}{s^3 + 6s^2 + 11s + 6} \; .$$

Determine the gain margin and hence the maximum value of K for a stable response.

$$\Phi_o(j\omega) = \frac{K}{(6 - 6\omega^2) + j\omega(11 - \omega^2)} \; .$$

$$\text{Thus Re } \Phi_o(j\omega) = \frac{6K(1 - \omega^2)}{\omega^6 + 14\omega^4 + 49\omega^2 + 36}$$

$$\text{Im } \Phi_o(j\omega) = \frac{-K\omega(11 - \omega^2)}{\omega^6 + 14\omega^4 + 49\omega^2 + 36} \; .$$

Hence the following table.

ω	Re $\Phi_o(j\omega)$	Im $\Phi_o(j\omega)$
$-\infty$	0	0
– ve large	– ve small	– ve small
$-\sqrt{11}$	– ve	0
– ve small	– ve	+ ve
-1	0	+ ve
0	$k/6$	0

From this table the Nyquist diagram can be sketched as shown:

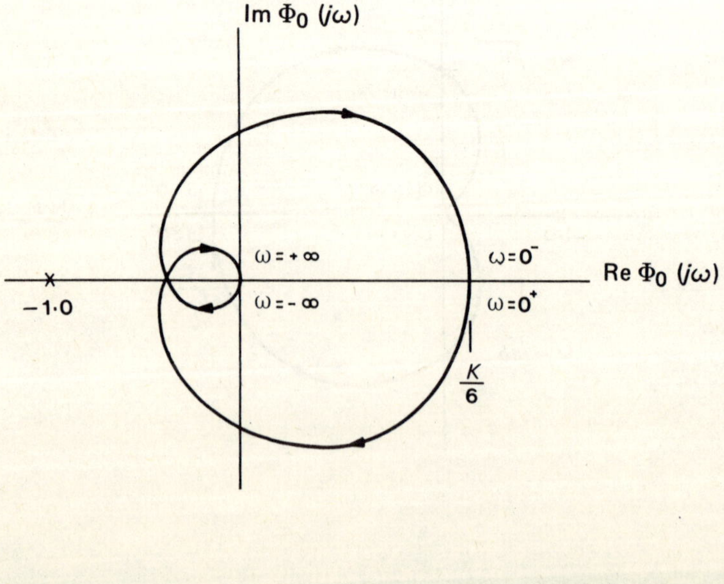

When $\omega = -\sqrt{11}$ $\text{Im}\,\Phi_o(j\omega) = 0$

and $$\text{Re}\,\Phi_o(j\omega) = \frac{6K(1-11)}{11^3 + 14.11^2 + 49.11 + 36} = -\frac{K}{60},$$

$$\text{i.e. the gain margin} = \frac{60}{K}$$

and K can be increased to 60 before instability occurs.

Alternatively, the maximum value of K for stability can be found by putting $\text{Re}\,\Phi_o(j\omega) = -1$ when $\text{Im}\,\Phi_o(j\omega) = 0$. In this case $-1 = -K/60$, and $K = 60$, as before.

CHAPTER 6

Exercises

6.1 SYSTEMS HAVING ONE DEGREE OF FREEDOM

1. The figure shows a pendulum which consists of a light rigid rod of length L pivoted to a fixed point at one end and having a mass m fixed to its other end. A spring of stiffness k is attached as shown, at a distance a from the pivot. In the position shown the rod is vertical and the spring is horizontal and unloaded. Find the frequency of free oscillations of small amplitude in the plane of the diagram.

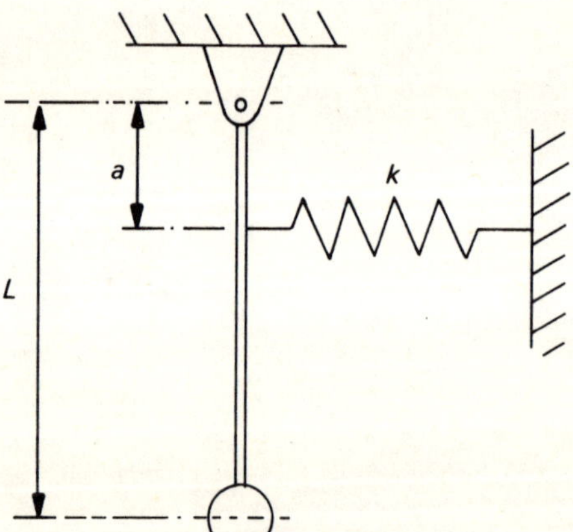

$$\left(\frac{1}{2\pi}\sqrt{\left(\frac{mgL + ka^2}{mL^2}\right)}\text{Hz}\right)$$

2. A uniform wheel of radius R can roll without slipping on an inclined plane. Concentric with the wheel and fixed to it is a drum of radius r aroung which is wrapped one end of string. The other end of the string is fastened to an anchored spring, of stiffness k, as shown. Both spring and strings are parallel

to the plane. The total mass of the wheel/drum assembly is m and its moment of inertia about the axis through the centre of the wheel O is I. If the wheel is displaced a small distance from its equilibrium position and released, derive the equation describing the ensuing motion and hence calculate the frequency of the oscillations. Damping is negligible.

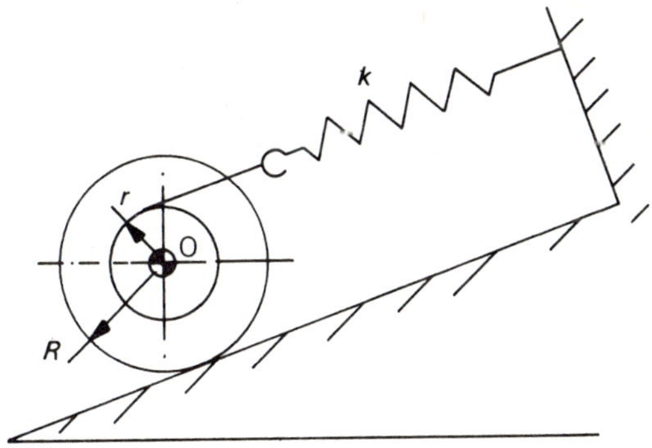

$$\left(\frac{1}{2\pi}\sqrt{\left(\frac{k(R+r)^2}{I+mR^2}\right)}\ \text{Hz}\right)$$

3. A wire is stretched between two fixed horizontal supports a distance L apart by a force T. A body of mass m is attached to the wire at a fixed distance x from one support. Find the natural frequency of small transverse vibrations of the body, assuming that T is constant.

$$\left(\frac{1}{2\pi}\sqrt{\left(\frac{TL}{mx(L-x)}\right)}\ \text{Hz}\right)$$

4. Find the natural frequency of free vibration for the system shown, when the amplitude of vibration is 2 cm. The mass of the body is 10 kg and each spring has a constant stiffness of 10 kN/m.

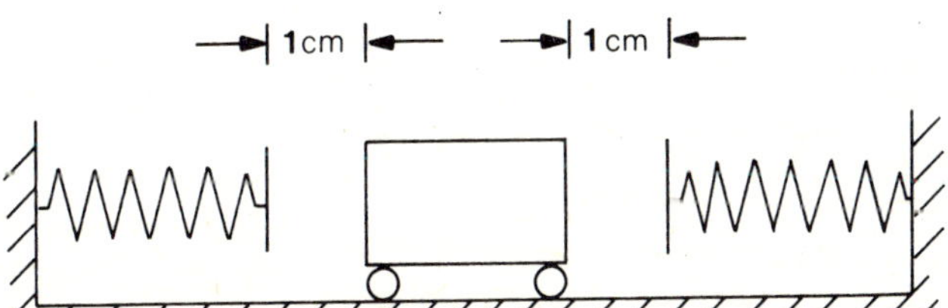

5. A body of mass 10 kg oscillates freely in a horizontal direction under the action of a spring for which the force-deflection curve is shown. Neglect damping and calculate the natural frequency of the system when the body is released from rest with a deflection of 2 cm.

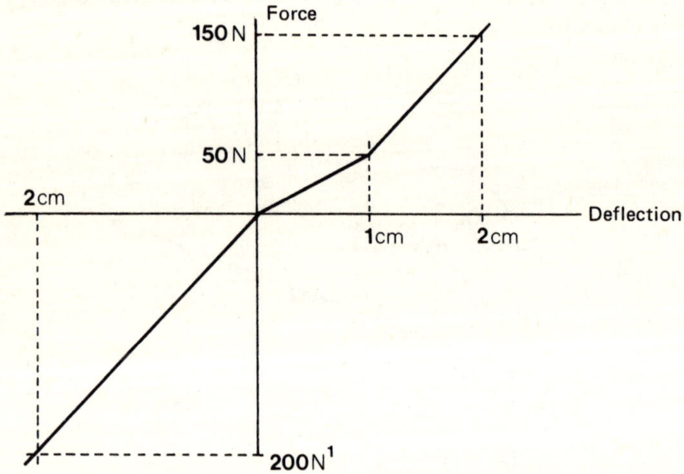

6. A light rigid rod of length L is pinned at one end **O** and has a body of mass m attached at the other end. A spring and viscous damper connected in parallel are fastened to the rod at a distance a from the support. The system is set up in a horizontal plane: a plan view is shown.
If the damper is adjusted to provide critical damping, obtain the motion of the rod as a function of time if it is rotated through a small angle θ_o and then released. If θ_o = 2 deg and the undamped natural frequency of the system is 3 rad/s, calculate the displacement 1 s after release.
Explain the term *logarithmic decrement* as applied to such a system and calculate its value assuming that the damping is reduced to 80% of its critical value.

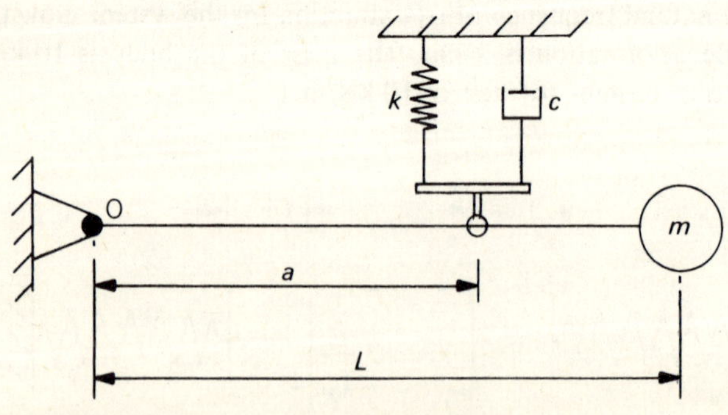

(0.4°, 8.38)

7. A spring of stiffness k is connected to a viscous damper in series, and released from rest with a displacement x_o from the zero force position. Neglect the mass of the system and express the displacement x as a function of the time t after release.

8. A body supported by an elastic structure performs a damped oscillation of period 1 s, in a medium whose resistance is proportional to the velocity. At a given instant the amplitude was observed to be 100 mm and in 10 s this had reduced to 1 mm.
   What would be the period of the free vibration if the resistance of the medium were negligible?

   (0.997 s)

9. A swing door is a uniform rectangular slab 0.9 m wide and mass 17 kg. It is pivoted about one vertical edge to a device which provides a spring moment proportional to the angle of the door from its closed position, acting in parallel with a viscous damper. When the door is held open at 90°, a moment of 16 Nm is needed to keep it in that position, and when it is released from that position it swings through the closed position 2.3 seconds later.
   Find to an accuracy of 5% or better, the damping ratio of the system.

   (0.72)

10. A turbine disc of mass 20 kg has its mass centre on the axis of a central hole 40 mm in diameter. The disc is suspended from a horizontal knife edge passed through the central hole, and makes small free oscillations in the plane of the disc while immersed in a fluid which exerts a resistance proportional to the angular velocity of the disc.
    The period of the oscillations is found to be 1.6 s and the amplitude is observed to decrease sixfold in each complete oscillation. Write down the equation of motion and derive its solution, and hence calculate the mass moment of inertia of the disc about the axis of the central hole and the effective viscous damping coefficient.

    (0.228 kgm^2, 0.0508 Nms/rad)

11. The moving part of a galvanometer has a mass moment of inertia I of 1.5×10^{-6} kg m^2 and is connected to the frame by a spiral spring which has an effective stiffness k of 1.2×10^{-5} Nm/rad and a viscous damper of coefficient c. Write down the equation of motion and derive from it the value of c which will make the movement just not oscillatory.

    (8.48×10^{-6} Nms/rad)

12. An indicator is operated by a solid cylindrical float of mass m, diameter d and length D, which moves with its axis vertical in a coaxial vertical tube of diameter D, as shown. The float is connected by a rack and pinion to a rotating indicator disc, which can be treated as a uniform disc of mass m and radius r: r is also the pitch circle radius of the pinion. The link between the float and the rack is of negligible mass, as is the rack. In static equilibrium the float is half immersed in the liquid in the tube.
Assuming that no liquid enters or leaves the tube and that viscous forces are negligible, find the frequency of small free oscillations of the system.

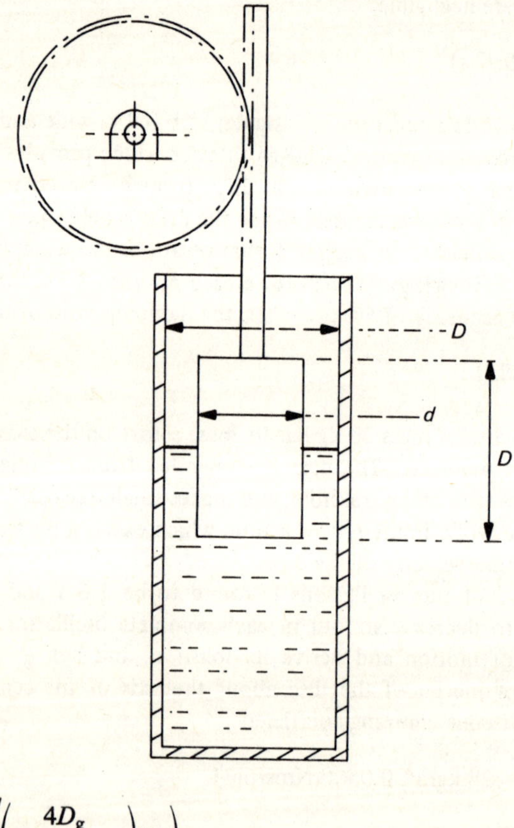

$$\left(\frac{1}{2\pi}\sqrt{\left(\frac{4Dg}{3(D^2-d^2)}\right)}\text{Hz}\right)$$

13. A uniform door of width 0.8 m and mass 20 kg is closed by a torsion spring of stiffness 100 Nm/rad and a viscous damper of coefficient c. Write down the equation of motion of the door whilst closing, give the solution, and

comment on the response of the door for amounts of damping less than, equal to and greater than critical. Find the value of c for critical damping. With the damping made critical, the door is held open at 90° and released; find, to the nearest tenth of a second, the time taken for the door to turn through 89°.

(1.3 s)

14. A U-tube manometer with two vertical arms has an internal diameter of 6mm, and contains mercury (density 13.6 g/cm^3), the total length of the mercury column being 0.8 m. Assuming the mercury to move as an incompressible body, friction between the mercury and the inside of the tube can be expressed as a shear force c per unit wetted area per unit velocity.
When the mercury column is displaced from its equilibrium position and then allowed to move freely, a damped oscillation is observed, the amplitude at the end of the eleventh cycle being one tenth of the amplitude at the end of the first cycle. Write the equation of motion, and hence find the value of c (in Ns/m^3) and the natural frequency of the oscillation.

(7.9, 0.789 Hz)

15. An aircraft ejector seat can be regarded as a horizontal platform which can be given a vertical acceleration, and the pilot can be treated (greatly simplified) as a rigid mass of 90 kg supported on the platform by springs of total effective stiffness 10.9 kN/m. When the pilot is to be ejected, the platform is given a constant vertically upward acceleration of 49 m/s^2 which lasts for 0.19 s, when the pilot is freed from the platform and continues to move through the air.
Considering only the vertical component of motion and neglecting the effect of air resistance find
(a) the maximum height to which the pilot rises above his original position of static equilibrium, and
(b) the maximum acceleration to which he is subjected.

(1.78 m, 73.5 m/s^2)

16. The natural frequency of the undamped spring-body system of the vibrometer shown 5 Hz. When the vibrometer is fixed to a horizontal surface of a motor running at 750 rev/min, the absolute amplitude of vibration of the suspended body is observed to be 1 mm. What is the amplitude of vertical vibration of the motor, and at what motor speed will the amplitude of motor vibration be the same as the amplitude of vibration of the vibrometer body?

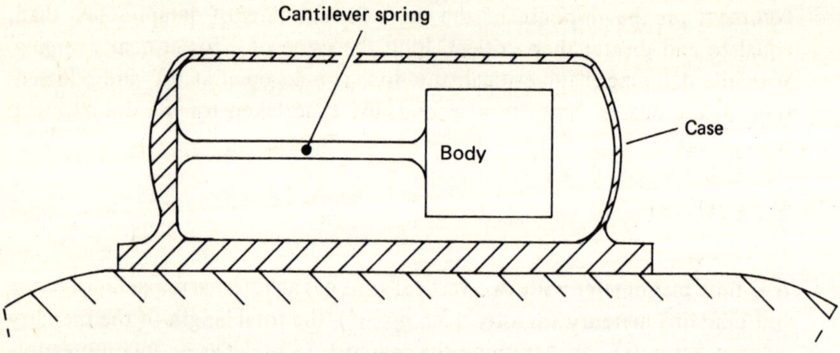

(5.25 mm, 424 rev/min)

17. The figure shows a vibration exciter which consists of two contra-rotating wheels, each carrying an eccentric body of mass 0.1 kg of an effective radius of 10 mm from its axis of rotation. The vertical positions of the eccentric bodies occur at the same instant. The total mass of the exciter is 2 kg and damping is negligible.
Find a value for the stiffness of the spring mounting so that a force of amplitude 100 N is transmitted to the fixed support due to rotor unbalance when the wheels rotate at 150 rad/s.

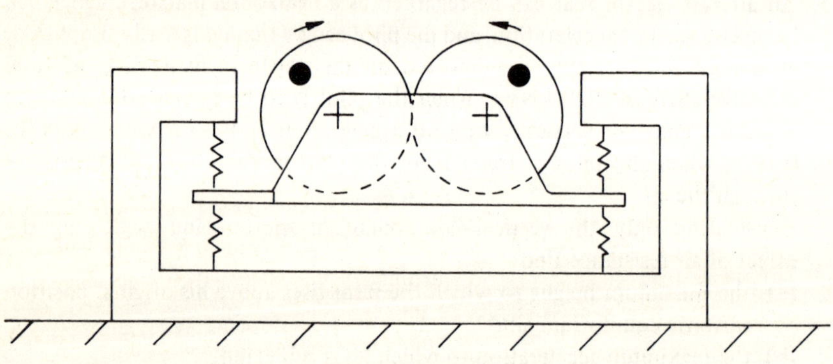

(81.92 kN/m, 31 kN/m)

18. A motor generator set of total mass 365 kg is mounted on damped vibration isolators. It runs at 1450 rev/min, and the unit is observed to have a vibration amplitude of 0.76 mm. When the unit is disturbed at standstill, it is found that the resulting damped oscillations have a frequency of 23 Hz and a decay factor of 1.5. Find

(a) The coefficient of viscous damping,
(b) The magnitude of the exciting force at the normal running speed,
and (c) The amplitude of the forced vibration if damping were not present.

(6750 N/m/s, 1000 N, 1.22 mm)

19. A piston is made to reciprocate in a vertical cylinder with absolute motion $X_p \cos \nu t$. The cylinder, of mass 2 kg, is attached to a platform of negligible mass and supported by springs of total stiffness 5 kN/m. If the viscous drag between the piston and cylinder is characterised by the coefficient $c = 200$ N/m/s, find the amplitude X_c of the absolute motion of the cylinder, and its phase angle with respect to the piston motion.

($X_c = 0.02$ m; X_c lags X_p by $\tan^{-1} 3/4$)

20. A machine of mass 520 kg produces a vertical disturbing force which oscillates sinusoidally at a frequency of 25 Hz. The force transmitted to the floor is to have an amplitude, at this frequency, not more than 0.4 times that of the disturbing force in the machine, and the static deflection of the machine on its mountings is to be as small as possible consistent with this. For this purpose, rubber mountings are to be used, which are available as units, each of which has a stiffness of 359 kN/m and a damping coefficient of 2410 Ns/m. How many of these units are needed?

(8)

21. A fan is mounted on a spring and viscous damper in parallel so that only linear motion in the vertical direction occurs. Briefly derive an expression for the force transmitted to the ground through the spring and damper if the fan generates a harmonic disturbing force in the vertical direction.
If the fan has a mass of 40 kg and a rotating unbalance of 0.01 kg m, determine the spring stiffness required for 10% force transmission. Take the damping ratio of the system to be 0.2 and the fan speed as 1480 rev/min.
When running under these conditions what effect on the transmission would removal of the damping element have?

(43 kN/m, 0.0469)

22. The basic element of many vibration-measuring devices is the seismic unit, which consists of a mass m supported from a frame by a spring of stiffness k in parallel with a damper of viscous damping coefficient c. The frame of

the unit is attached to the structure whose vibration is to be determined, the quantity measured being z, the relative motion between the seismic mass and the frame. The motion of both the structure and the seismic mass is translation in the vertical direction only.

Derive the equation of motion of the seismic mass, assuming that the structure has simple harmonic motion of circular frequency υ, and deduce the steady state amplitude of z.

If the undamped natural circular frequency ω of the unit is much greater than υ, show why the unit may be used to measure the acceleration of the structure. Explain why in practice some damping is desirable.

If the sensitivity of the unit (i.e. the amplitude of z as a multiple of the amplitude of the acceleration of the frame) is to have the same value when $\upsilon = 0.2\ \omega$ as when $\upsilon \ll \omega$, find the necessary value of the damping ratio.

(0.7)

23. A machine of mass 940 kg produces a vertical disturbing force which oscillates sinusoidally, having an amplitude of 320 N at the machine's normal running speed of 550 rev/min. At this speed the force transmitted to the floor is to have an amplitude of not more than 100 N, and to achieve this the machine is to be mounted on viscously damped springs. The static deflection of the spring mountings is to be not more than 15 mm, and the amplitude of vibration at resonance is to be kept as low as possible. Find (a) the total spring stiffness and total viscous damping coefficient which will enable these conditions to be satisfied, and (b) the resonant speed of the machine when such mountings are fitted.

(615 kN/m, 8.85 kN/m/s)

24. The vibration of the floor of a laboratory is found to be simple harmonic motion at a frequency in the range 15 – 60 Hz (depending on the speed of some nearby reciprocating plant). It is desired to install in the laboratory a sensitive instrument which requires insulating from the floor vibration. The instrument is to be mounted on a small platform which is supported by three similar springs resting on the floor, arranged to carry equal loads; the motion is restrained to occur in a vertical direction only. The combined mass of the instrument and the platform is 40 kg: the mass of the springs can be neglected and the equivalent viscous damping ratio of the suspension is 0.2.

Calculate the maximum value for the spring stiffness if the amplitude of the transmitted vibration is to be less than 10% of that of the floor vibration, over the given frequency range.

(5.3 kN/m)

25. A galvanometer consists of a coil and mirror assembly suspended on a torsional spring in a fixed enclosure filled with a liquid which gives viscous damping. The moment of inertia of the moving part about its axis of rotation is 2×10^{-9} kg m^2, the spring stiffness is 0.0116 Nm/rad and the damping coefficient is 5.29×10^{-6} Nms/rad. Find the natural frequency of the galvanometer, and the amplitude and phase errors in its indication when supplied with a sinusoidal input signal at a frequency of 100 Hz and amplitude 100 mA.

(322 Hz, 2.5 mA, 17°)

26. The figure shows the essential elements of an instrument for recording vibrations, which consist of a frame A with a vertical pillar and light rod that pivots about a horizontal axis passing through the pillar at O. The rod has a concentrated mass m attached to its free end and is held in a horizontal position by a spring of stiffness K while its vibrations are damped by a viscous damper whose damping coefficient is C.
If the frame of the instrument is given a vertical displacement $a \sin \upsilon t$, consider the rotation of the rod about O and derive an equation for the maximum amplitude of the relative, steady-state, vertical displacements between the mass m and the frame in terms of the given quantities and the natural frequency of the free vibrations of the instrument.

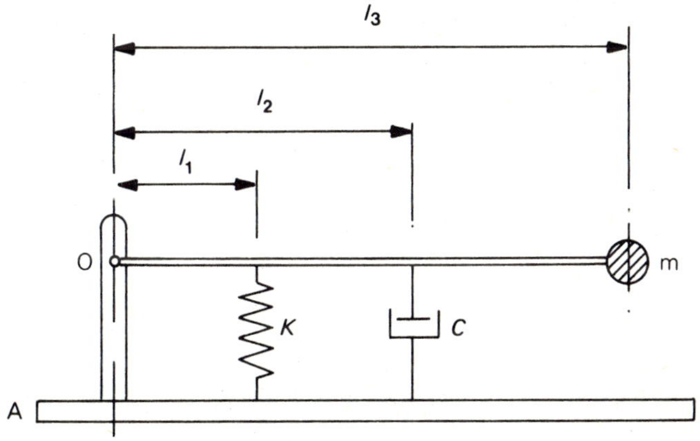

27. The figure shows, diagramatically, a trailer which is towed at a constant velocity v along a test track with a sinusoidal surface. The trailer has a mass M and is connected by springs of total stiffness k to the wheels which have a total mass m. The wavelength of the track surface is L and the amplitude of its undulations is h.

Derive an equation for the force exerted by the wheels on the track surface in terms of the given quantities, assuming that the only relative motion between the trailer and the wheels is in a vertical direction and that the diameter of the wheels is small compared with the undulations of the track.

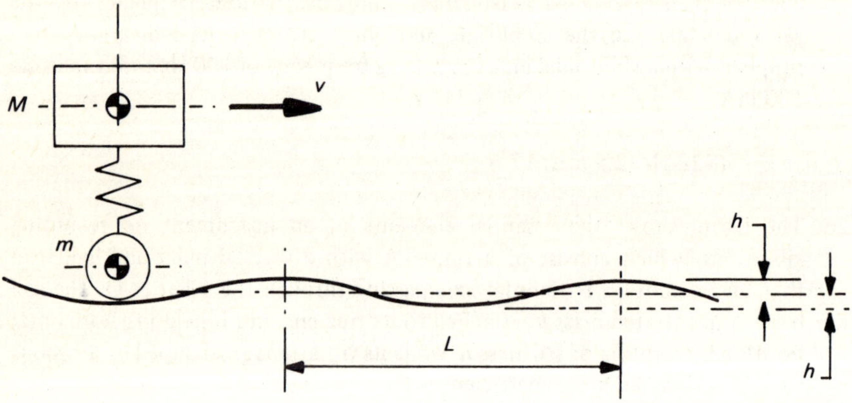

$$\left((M+m)g - \left[\frac{Mk}{k - M(2\pi v/L)^2} + m\right] h(2\pi v/L)^2 \sin(2\pi v/L)t)\right) .$$

28. A two-wheeled trailer of sprung mass 700 kg is towed at 60 km/h along an undulating straight road whose surface may be considered sinusoidal. The distance from peak to peak of the road surface is 30 m and the height from hollow to crest 0.1 m. The effective stiffness of the trailer suspension is 60 kN/m, and the shock absorbers which provide linear viscous damping are set to give a damping ratio of 0.67. Assuming that only vertical motion of the trailer is excited, find the absolute amplitude of this motion and its phase angle relative to the undulations.

   (0.056 m, 2.25°)

6.2 SYSTEMS HAVING MORE THAN ONE DEGREE OF FREEDOM

29. A turbine rotor with mass moment of inertia 12 kg m^2 is connected to a generator by a straight shaft of length 1 m and torsional stiffness 600 kNm/rad. The generator armature has a mass moment of inertia of 5 kg m^2. Calculate the natural frequency of free torsional oscillation of the system and give the mode shape. Ignore damping.

   (66 Hz, –2.4)

30. A tractor and a trailer have masses of 1000 kg and 500 kg respectively. They are connected by a coupling which consists of two springs in series, one spring having a stiffness of 200 kN/m and the other of 50 kN/m; there is negligible damping.
Find the frequency of free, small amplitude, longitudinal oscillations of the system when the tractor and trailer are in line and on a horizontal surface.

(1.74 Hz)

31. To analyse the vibrations of a two-storey building it is represented by the lumped mass system shown, where $m_1 = \frac{1}{2} m_2$ and $k_1 = \frac{1}{2} k_2$ (k_1 and k_2 represent the shear stiffnesses of the parts of the building shown).
Calculate the natural frequencies of free vibrations and sketch the corresponding mode shapes of the building, showing the amplitude ratios.
If a horizontal harmonic force $F_1 \sin \nu t$ is applied to the top floor, determine expressions for the amplitudes of the steady state vibration of each floor.

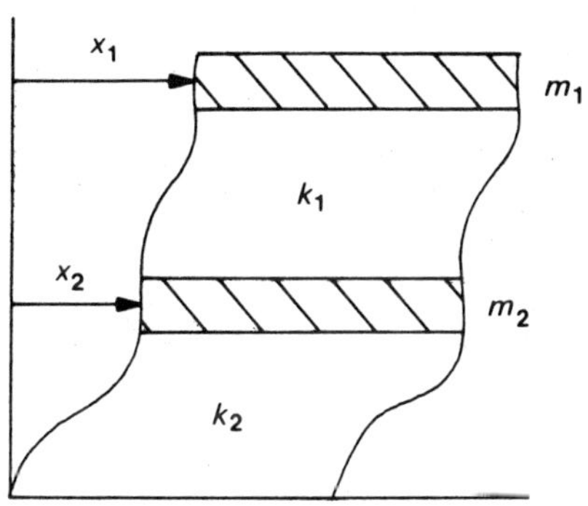

$$\left(\begin{array}{l} \dfrac{1}{2\pi}\sqrt{\left(\dfrac{k_1}{2m_1}\right)} \text{ Hz}, +0.5 \\ \dfrac{1}{2\pi}\sqrt{\left(\dfrac{2k_1}{m_1}\right)} \text{ Hz}, -1.0 \end{array} \right)$$

32. The rigid beam shown in its position of static equilibrium in the figure has a mass m and a mass moment of inertia $2ma^2$ about an axis perpendicular to the plane of the diagram and through its centre of gravity G. Assuming no horizontal motion of G, find the frequencies of small oscillations in the plane of the diagram and the corresponding positions of the nodes.

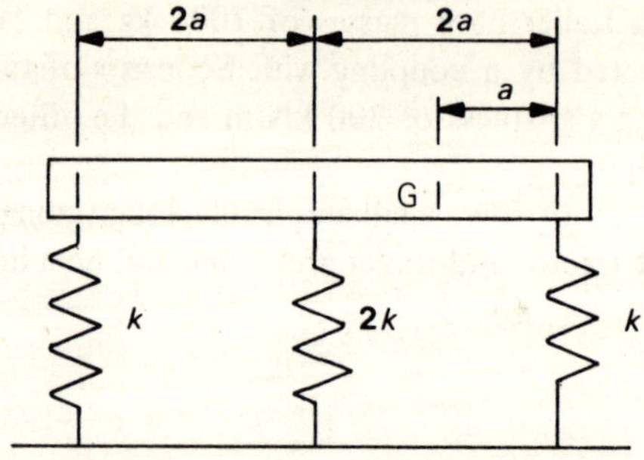

$$\left(\frac{1}{2\pi}\sqrt{(2k/m)}\ \text{Hz}, \frac{1}{2\pi}\sqrt{(8k/m)}\ \text{Hz}\right)$$

33. The figure shows a car mounted on its springs which are constrained so that the centre of gravity of the car G can only move vertically. The car has a mass m and a radius of gyration K_0 about an axis through G normal to the plane of the figure. The real springs have a total stiffness k_1 and are at a distance a from G and the front springs have a total stiffness k_2 and are at a distance b from G.
Derive an equation for the natural frequencies of the vibrations of the car in terms of the natural frequencies of pure vertical, or bounce, vibrations and of pure rotational, or pitch, vibrations and the other parameters of the system.
Show that when there is no coupling between the bounce and pitch modes of vibration, the natural frequencies in bounce and pitch are equal.

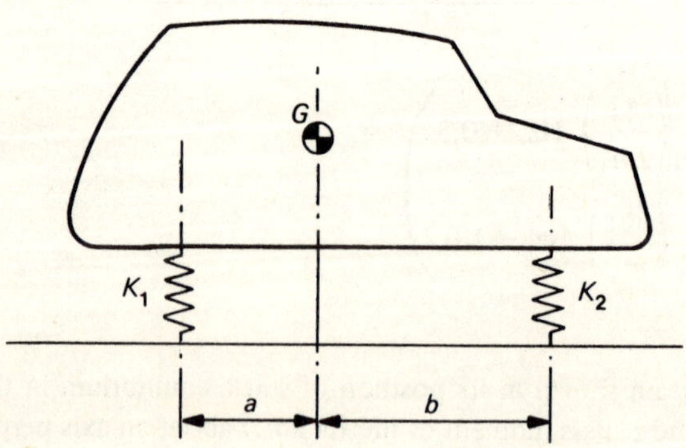

$$(\omega^4 - \omega^2(\omega_B^2 + \omega_p^2) + \omega_B^2\omega_p^2 - (k_2 b - k_1 a)^2/m^2 K_0^2 = 0)$$

34. A film reel is driven by a tightly wound helical spring stretched round two grooved pulleys, as shown. The pulleys 1 and 2 have radii r_1 and r_2, angular displacements θ_1 and θ_2, and moments of inertia about their axes of rotation I_1 and I_2, respectively. The stiffness of each unsupported section of the spring is k.
Derive an expression for the natural frequency of small amplitude free oscillations of the system. Assume that there is no damping and note that during oscillation $r_1\dot{\theta}_1 \neq r_2\dot{\theta}_2$.

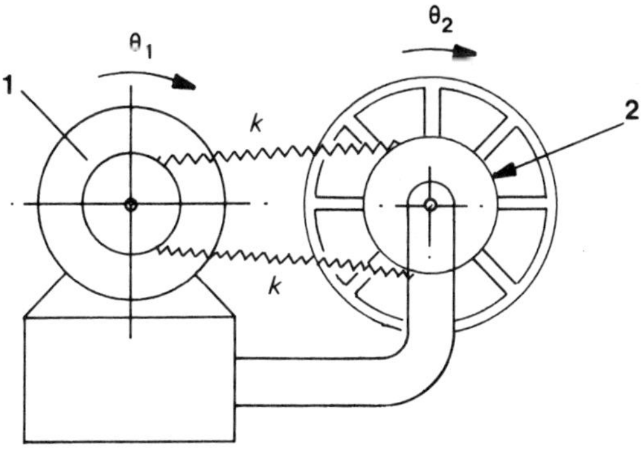

$$\left(\frac{1}{2\pi}\sqrt{\left(2k\,\frac{I_1 r_2^2 + I_2 r_1^2}{I_1 I_2}\right)}\text{ Hz}\right)$$

35. A small electronic package is supported by springs as shown. The mass of the package is m each spring has a constant axial stiffness k and damping is negligible. Considering motion in the plane of the figure only, and assuming that the amplitude of vibration of the package is small enough for the lateral spring forces to be negligible, write down the equations of motion and hence obtain the frequencies of free vibration of the package.
Explain how the vibration mode shapes can be found.

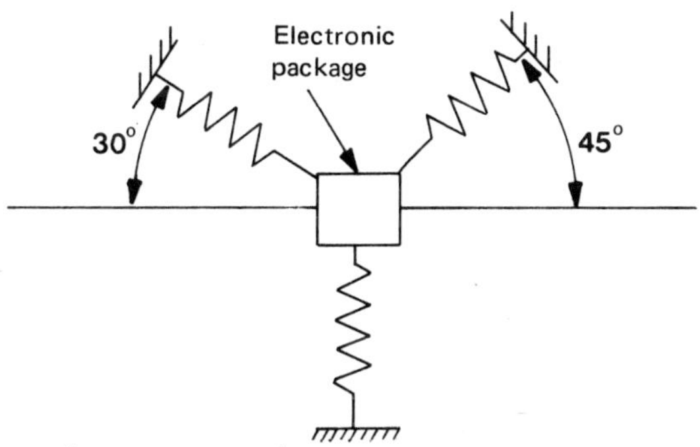

$$\left(\frac{1.11}{2\pi}\sqrt{\left(\frac{k}{m}\right)}\text{ Hz},\ \frac{1.33}{2\pi}\sqrt{\left(\frac{k}{m}\right)}\text{ Hz}\right)$$

36. Explain, in one sentence each, what is meant by a *natural frequency* and *mode shape* of a dynamic system.

Part of a machine can be modelled by the system shown. Two uniform discs A and B, which are free to rotate about fixed parallel axes through their centres, are coupled by a spring. Similar springs connect the discs to the fixed frame as shown in the figure. Each of the springs has a stiffness of 2.5 kN/m which is the same in tension or compression. Disc A has a mass moment of inertia about its axis of rotation of 0.05 kg m^2, and a radius of 0.1 m, whilst for the disc B the corresponding figures are 0.3 kg m^2 and 0.2 m. Damping is negligible.

Determine the natural frequencies of small amplitude oscillation of the system and the corresponding mode shapes.

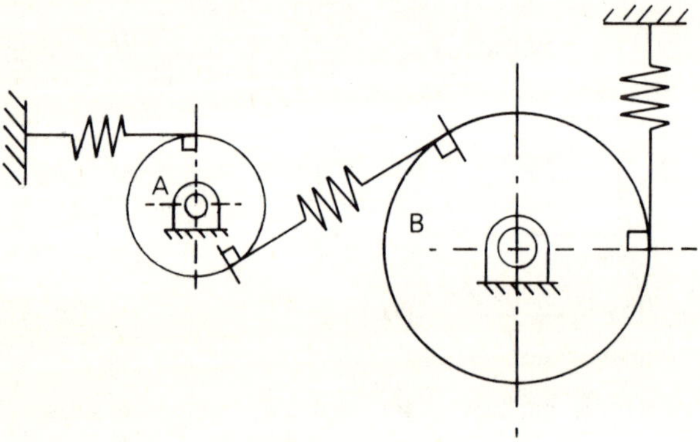

(3.17 Hz, 5.68 Hz, 3.68, –1.65)

37. Briefly derive the equations which describe the operation of an undamped dynamic vibration absorber.

A milling machine of mass 2700 kg demonstrates a large resonant vibration in the vertical direction at a cutter speed of 300 rev/min when fitted with a cutter having 20 teeth. To overcome this effect it is proposed to add an undamped vibration absorber.

Calculate the minimum absorber mass and the relevant spring stiffness required if the resonant frequency is to lie outside the range corresponding to a cutter speed of 250 to 350 rev/min.

(363 kg, 135 MN/m)

38. In a pumping station, a section of pipe resonated at a pump speed of 120 rev/min. To eliminate this vibration, it was proposed to clamp a spring-mass system to the pipe to act as an absorber. In the first test, an absorber mass of 2 kg tuned to 120 cycle/min resulted in the system having a natural frequency of 96 cycle/min.

If the absorber system is to be designed so that the natural frequencies lie outside the region 85 – 160 cycle/min, what are the limiting values of the absorber mass and spring stiffness?

(4.9 kg, 773 N/m)

39. A certain machine of mass 300 kg produces an harmonic disturbing force $F \cos 15t$. Because the frequency of this force coincides with the natural frequency of the machine on its spring mounting an undamped vibration absorber is to be fitted.
    If no resonance is to be within 10% of the exciting frequency, find the minimum mass and corresponding stiffness of a suitable absorber. Derive your analysis from the equations of motion, treating the problem as one-dimensional.

40. A machine tool of mass 3000 kg has a large resonant vibration in the vertical direction at 120 Hz. To control this resonance, an undamped vibration absorber of mass 600 kg is fitted, tuned to 120 Hz. Find the frequency range in which the amplitude of the machine vibration is less with the absorber fitted than without.

(102 – 140 Hz)

41. The figure shows a body of mass m_1 which is supported by a spring of stiffness k_1 and which is excited by an harmonic force $P \sin vt$. An undamped dynamic vibration absorber consisting of a mass m_2 and a spring of stiffness k_2 is attached to the body as shown.

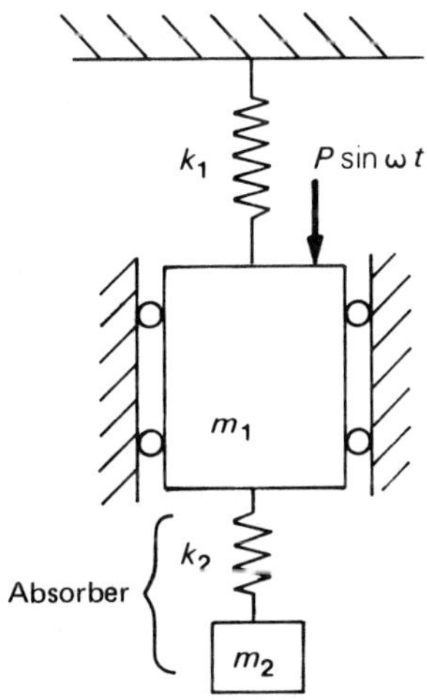

Derive an expression for the amplitude of the vibrations of the body.
The body shows a violent resonance at 152 Hz. As a trial remedy a vibration absorber is attached which results in a resonant frequency at 140 Hz. How many such absorbers are required if no resonance is to occur between 120 and 180 Hz?

42. The figure shows a steel shaft which can rotate in frictionless bearings, and which has three discs rigidly attached to it in the positions shown. Each of the three discs is solid, uniform, and of diameter 280 mm. The Modulus of Rigidity of the shaft material is 83 GN/m^2, and the density of the discs is 7000 kg/m^3. Find the frequencies of free torsional vibrations of the system.

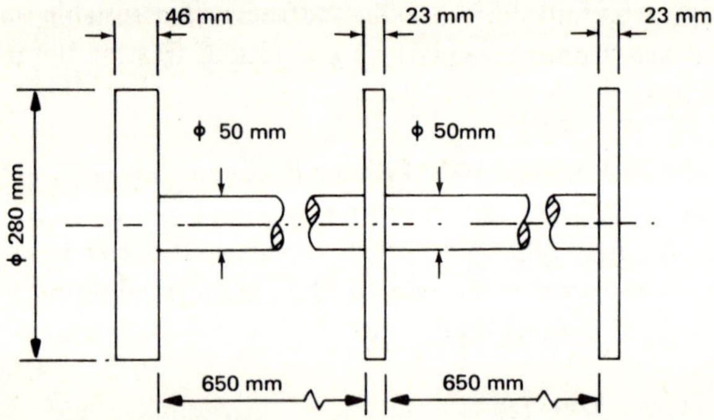

(121.6 Hz, 239 Hz)

43. An aeroplane has a fuselage mass of 4000 kg. Each wing has an engine of mass 500 kg, and a fuel tank of mass 200 kg at its tip, as shown. Neglecting the mass of each wing, calculate the frequencies of flexural vibration in a vertical plane. Take the stiffness of the wing sections to be $3k$ and k as shown, where $k = 100$ kN/m.

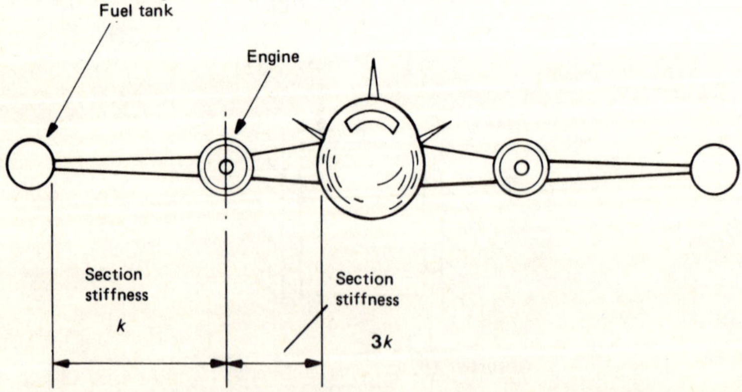

(3.09 Hz, 5.22 Hz)

44. A manufacturer of delicate equipment ships his products, which have a mass m, by fitting them with a spring mounting of stiffness k. It is found that damage to the equipment is caused by large amplitude vibrations in the vertical direction. A special packing is therefore considered. The packing case, of mass m, is supported on a spring mounting of stiffness k; two pieces of equipment, each of mass m, are supported within the case by three springs each of stiffness k, as shown (idealised).
Considering vibrations in the vertical direction only, find the natural frequencies of this system and hence comment on its suitability.

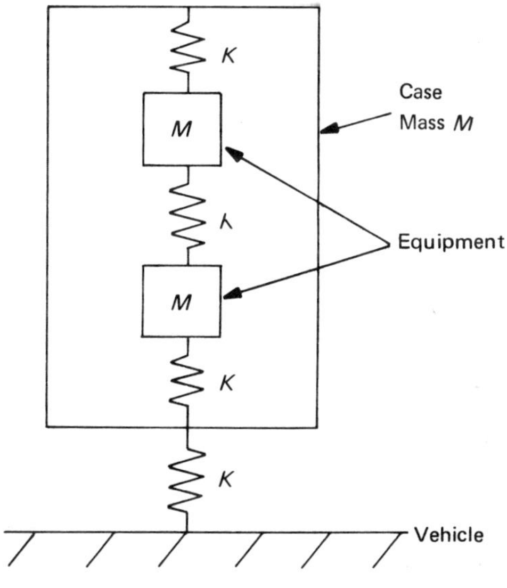

$$\left(\frac{0.52}{2\pi}\sqrt{\left(\frac{k}{m}\right)}, \frac{1.73}{2\pi}\sqrt{\left(\frac{k}{m}\right)}, \frac{1.93}{2\pi}\sqrt{\left(\frac{k}{m}\right)}\,\text{Hz}\right)$$

45. A marine propulsion installation is shown in the figure. For the analysis of torsional vibration, the installation can be modelled as the system shown, where the mass moments of inertia for the engine, gearbox and propeller taken about the axis of rotation are I_E, I_G and I_P respectively, and the stiffnesses of the gearbox and propeller shafts are k_G and k_P respectively. The numerical values are:

$$I_E = 0.8 \text{ kg m}^2$$
$$I_G = 0.3 \text{ kg m}^2$$
$$I_P = 2.0 \text{ kg m}^2$$
$$k_G = 400 \text{ kNm/rad}$$
$$k_P = 120 \text{ kNm/rad},$$

and damping can be neglected.

Calculate the natural frequencies of free torsional vibration and give the positions of the node or nodes for each frequency.

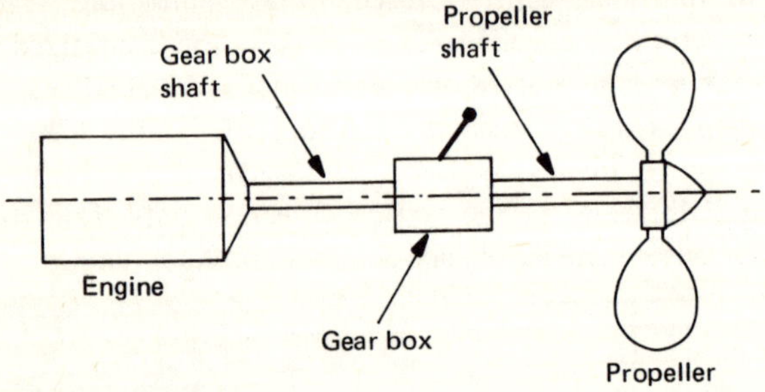

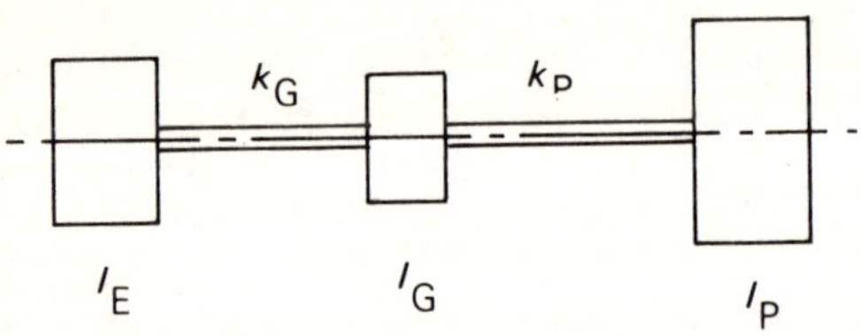

(60.5 Hz, + 1.4 – 0.697; 232 Hz, – 0.304 – 0.028)

46. A machine is modelled by the system shown. The masses of the main elements are m_1 and m_2 and the spring stiffnesses are as shown. Each roller has a mass m, diameter d, and mass moment of inertia J about its axis and rolls without slipping.
Considering motion in the longitudinal direction only, use Lagrange's equation to obtain the equations of motion for small free oscillations of the system. If $m_1 = 4m$, $m_2 = 2m$ and $J = md^2/8$, deduce the natural frequencies of the system and the corresponding mode shapes.

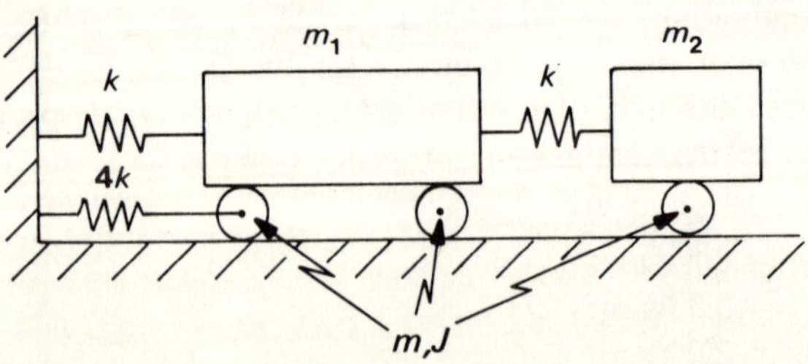

$$\left(\frac{0.459}{2\pi}\sqrt{\left(\frac{k}{m}\right)}\text{ Hz}, +0.5;\ \frac{0.918}{2\pi}\sqrt{\left(\frac{k}{m}\right)}\text{ Hz}, -1.0\right)$$

47. A simplified model for studying the dynamics of a motor vehicle is shown. The body has a mass M and a moment of inertia about an axis through its mass centre of I_G. It is considered to be free to move in two directions, vertical translation and rotation in the vertical plane. Each of the unsprung wheel masses, m, are free to move in vertical translation only. Damping effects are ignored.
    (a) Derive equations of motion for this system. Define carefully the co-ordinates used.
    (b) Is it possible to determine the natural frequency of a 'wheel hop' mode without solving all the equations of motion? If not, suggest an approximation which might be made in order to obtain an estimate of the wheel hop frequency and calculate such an estimate given the following data:

    $k = 20$ kN/m; $K = 70$ kN/m; $m = 22.5$ kg.

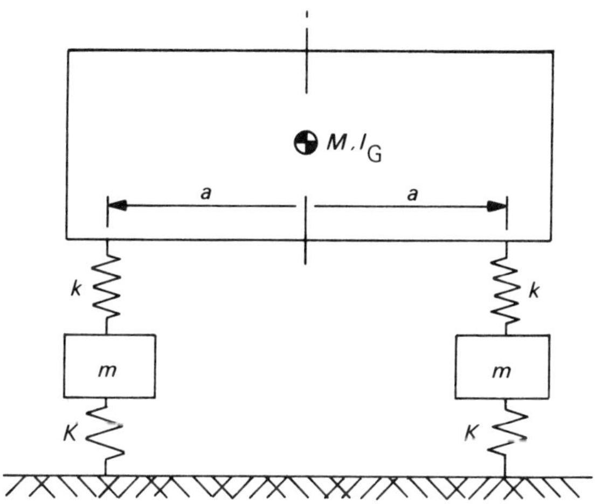

(10.1 Hz)

48. To analyse the vibration of a two-coach rail unit, it is modelled as the system shown. Each coach is represented by a rigid uniform beam of length l and mass m: the coupling is a simple ball-joint. The suspension is considered to be three similar springs, each of stiffness k, positioned as shown. Damping can be neglected.

    Considering motion in the plane of the figure only, obtain the equations of motion for small amplitude free vibrations and hence obtain the natural frequencies of the system.

Explain how the mode shapes may be found.

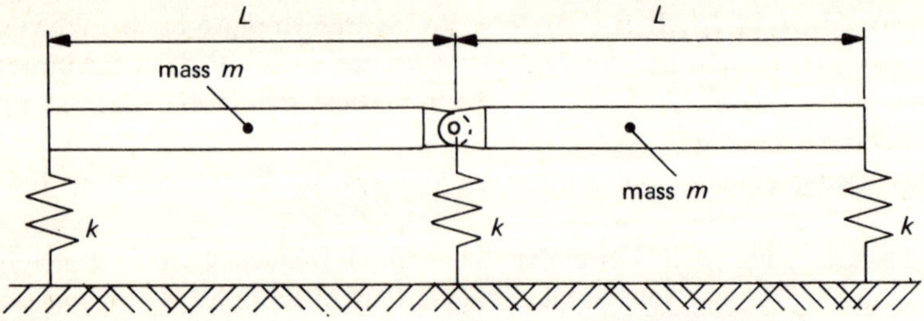

$$\left(\frac{1}{2\pi}\sqrt{\left(\frac{3k}{m}-\frac{\sqrt{3}k}{m}\right)}\ \text{Hz},\ \frac{1}{2\pi}\sqrt{\left(\frac{3k}{m}\right)}\ \text{Hz},\ \frac{1}{2\pi}\sqrt{\left(\frac{3k}{m}+\frac{\sqrt{3}k}{m}\right)}\ \text{Hz}\right)$$

49. A structure is modelled by three identical long beams and rigid bodies, connected by two springs as shown.
The rigid bodies are each of mass M and the mass of the beams is negligible. Each beam has a transverse stiffness K at its unsupported end, and the springs have stiffnesses k and $2k$ as shown.
Determine the frequencies and corresponding mode shapes of small amplitude oscillation of the bodies in the plane of the figure. Damping can be neglected.

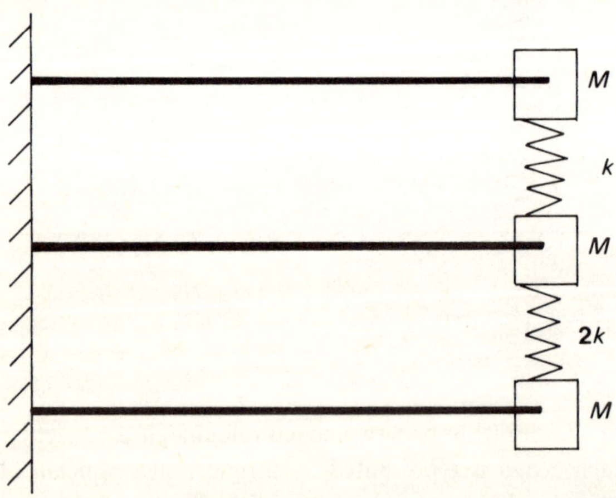

50. A tractor and trailer, each of mass M, are shown idealised in the figure. The trailer carries a heavy cylindrical drum of mass $\frac{2}{3}M$ and radius r; the drum can roll without slipping on the trailer. Both the tractor/trailer coupling and the trailer/drum attachment have a stiffness K.

Considering motion in the longitudinal (fore and aft) direction only, use Lagrange's equation to obtain the equations of motion for small free oscillations of the system. Hence deduce the natural frequencies of the system and the corresponding mode shapes.

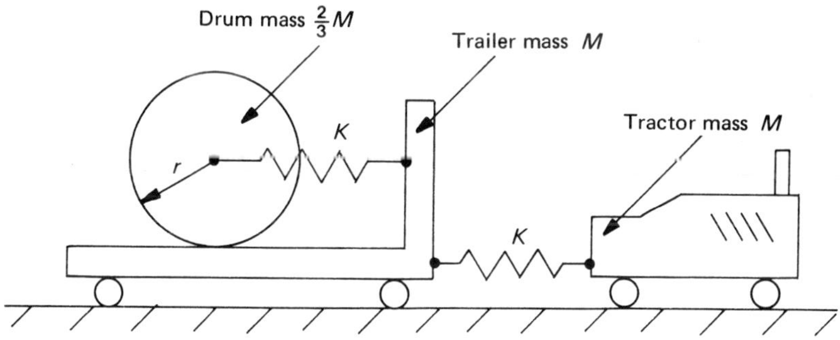

$$\left(0 \text{ Hz}, 1{:}1{:}1;\ \frac{1}{2\pi}\sqrt{\left(\frac{k}{M}\right)} \text{ Hz}, 1{:}0{:}-1.5;\ \frac{1}{2\pi}\sqrt{\left(\frac{24k}{11M}\right)} \text{ Hz}, 1{:}-\frac{13}{11}{:}\frac{3}{11}\right)$$

51. A system is modelled as three identical simple pendula connected by two similar springs as shown.
Each pendulum is freely pivoted, has a length l and a concentrated mass m: each spring has a stiffness k. The spring attachments to each pendulum are at a distance h from the pivot.
Determine the frequencies and corresponding mode shapes of small amplitude oscillation of the pendula in the plane of the figure.

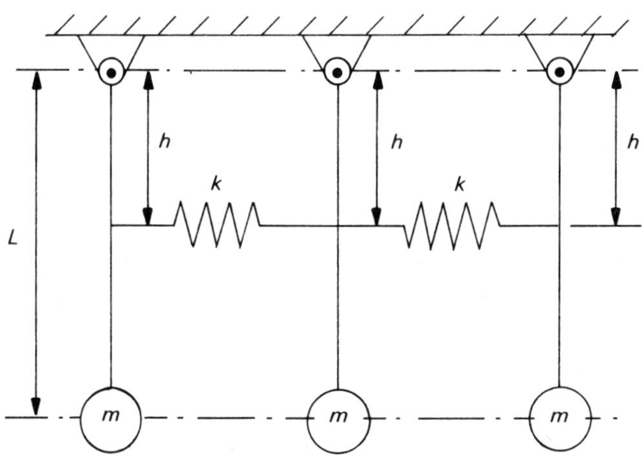

$$\left(\begin{array}{l} \dfrac{1}{2\pi}\sqrt{\left(\dfrac{g}{l}\right)}\ \text{Hz, } 1{:}1{:}1 \\ \dfrac{1}{2\pi}\sqrt{\left(\dfrac{g}{l}+\dfrac{kh^2}{ml^2}\right)}\ \text{Hz, } 1{:}0{:}-1 \\ \dfrac{1}{2\pi}\sqrt{\left(\dfrac{g}{l}+\dfrac{3kh^2}{ml^2}\right)}\ \text{Hz, } 1{:}-2{:}1 \end{array}\right)$$

52. Vibrations of a particular structure can be analysed by considering the equivalent system shown.

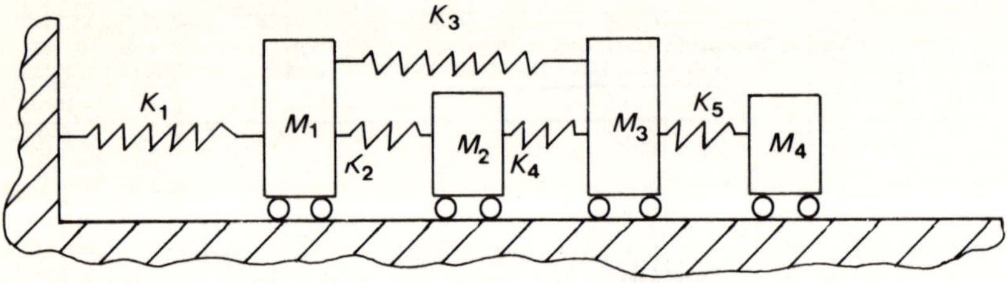

The bodies are mounted on small frictionless rollers whose mass is negligible, and motion occurs in a horizontal direction only.
Write down the equations of motion of the system and determine the frequency equation in determinant form.
Indicate how you would
(a) Solve the frequency equation.
(b) Determine the mode shapes associated with each natural frequency.
Briefly *describe* how the Lagrange equation could be used to obtain the natural frequencies of free vibration of the given system.

53. A simply supported beam of negligible mass and of length L has three bodies each of mass m attached as shown. The influence coefficients are, using standard notation

$$\alpha_{11} = 3L^3/256\,EI \qquad \alpha_{31} = 2.33L^3/256\,EI$$
$$\alpha_{21} = 3.67L^3/256\,EI \qquad \alpha_{22} = 5.33L^3/256\,EI$$

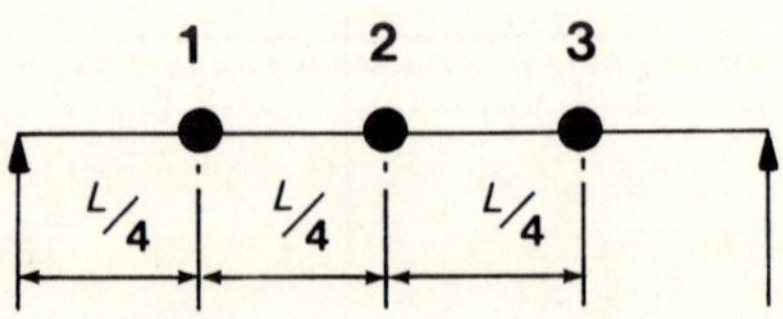

Write the flexibility matrix and determine by iteration the frequency of the first mode of vibration correct to 2 significant figures, if $EI = 10\ \text{Nm}^2$, $m = 2$ kg and $L = 1$ m.

Comment on the physical meaning of the eigenvector you have obtained and *describe* how you would use the orthogonality principle to obtain the frequencies of the higher modes.

54. Find the dynamic matrix of the system shown.

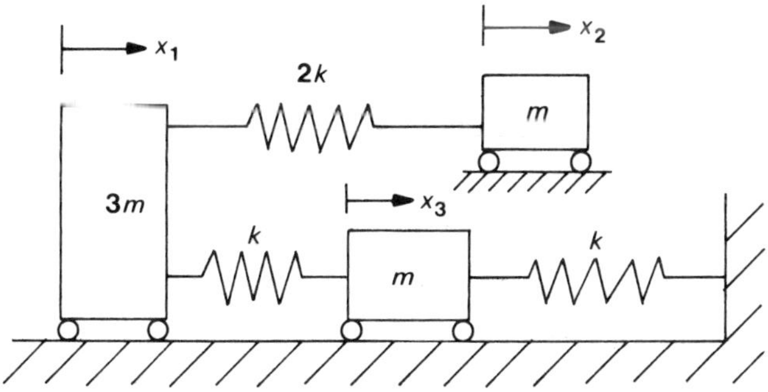

If $k = 20$ kN/m and $m = 5$ kg, find the lowest natural frequency of the system and the associated mode shape.

$$\left(\begin{bmatrix} k/m & -2k/3m & -k/3m \\ -2k/3m & 2k/3m & 0 \\ -k/m & 0 & 2k/3m \end{bmatrix}\right)$$

55. Define the term *degree of freedom* and explain the terms *normal mode of vibration* and *natural frequency* as applied to systems which have more than one degree of freedom.

56. Define the terms *direct receptance* and *cross receptance,* and explain how the receptance technique can be used in the analysis of complex dynamic systems.

57. Many dynamic systems excited by time-dependent quantities which are not harmonic, yet harmonic frequency response methods are sometimes used in the analysis of the response of such systems. How, and under what conditions, can this be justified?

58. All real mechanical systems posses damping and yet, in the vibration analysis of such systems, damping is often neglected.
Why is this done, and how is it justified?

59. An engine with a flywheel is shown. The engine is to drive a propeller of inertia I through a shaft of torsional stiffness k as indicated.

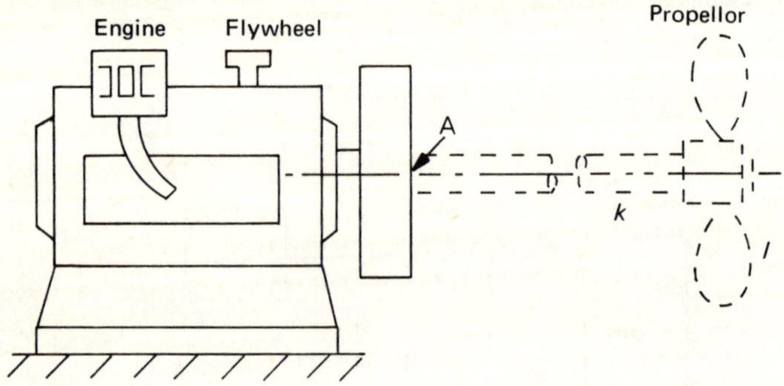

The receptance for torsional oscillation of the engine-flywheel system has been measured at point A over a limited frequency range which does not include any internal resonances of the system. The figure shows the receptance at A as a function of (frequency)^2.

(i) Calculate the lowest non-zero natural frequency for the engine, flywheel and propeller system if $I = 0.9$ kgm^2 and $k = 300$ kNm/rad.

(ii) Alternative designs propose to increase k or increase I. By considering the receptance of the shaft-propeller system, predict the effect of each of these proposals on the lowest non-zero natural frequency of the complete system.

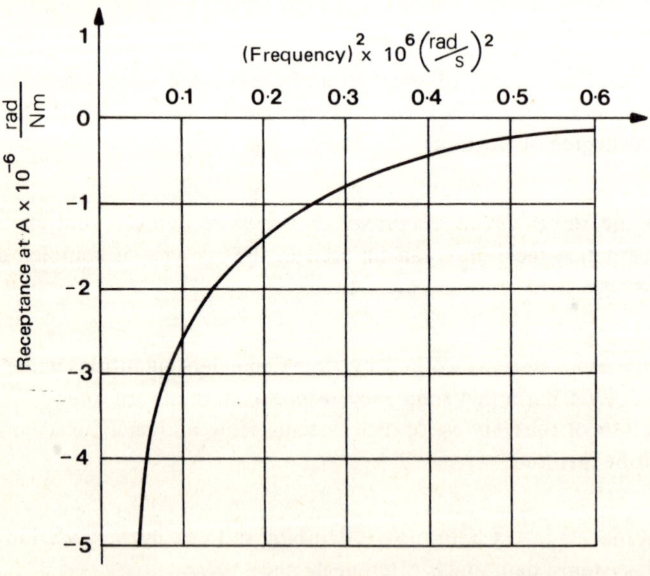

(99.5 Hz)

6.3 SYSTEMS WITH DISTRIBUTED MASS AND ELASTICITY

60. A uniform bar of length l is rigidly fixed at one end and is free at the other. Show that the frequencies of longitudinal vibration are

$$f = \frac{(n + \frac{1}{2})c}{2l} \text{ Hz}$$

where c is the velocity of longitudinal waves (E/ρ) in the bar, and $n = 0,1,2,\ldots$.

61. A uniform beam of length l is free at one end and has an axial force $F \sin vt$ applied at the other. Find an expression for the steady state vibration and hence deduce the natural frequencies of free vibration.

$$\left(F\left(\cos\left(\frac{v}{c}\right)x + \tan\left(\frac{v}{c}\right)L.\ \sin\left(\frac{v}{c}\right)x\right)\sin vt\ .\right)$$

62. A solid steel shaft, 25 mm in diameter and 0.45 m long, is mounted in long bearings in a rigid frame at one end and has at its other end, which is unsupported, a steel flywheel. The flywheel can be treated as a rim 0.6 m in outer diameter and 20 mm square cross-section, with rigid spokes of negligible mass lying in the mid-plane of the rim.
Find the frequency of free flexural vibrations.

(19.4 Hz)

63. A uniform horizontal steel beam is built in to a rigid structure at one end and pinned at the other end; the pinned end cannot move vertically but is otherwise unconstrained. The beam is 8 m long, the relevant flexural second moment of area of the cross-section is 4.3×10^6 (mm)^4, and the beam's own mass together with the mass attached to the beam is equivalent to a uniformly distributed mass of 600 kg/m. Using a combination of sinusoidal functions for the deflected shape of the beam, estimate the lowest natural frequency of flexural vibrations in the vertical plane.

(1.64 Hz)

64. In the figure, the bearings at each end of the shaft allow the shaft to tilt freely; the shaft is uniform, of material having a Young's Modulus E and with a flexural second moment of area I_s. The two discs, each having a mass m and a diametral mass moment of inertia I, are rigidly attached to the shaft in the positions shown. Neglecting the mass of the shaft, estimate the lowest frequency of free flexural vibrations when the shaft is not rotating.

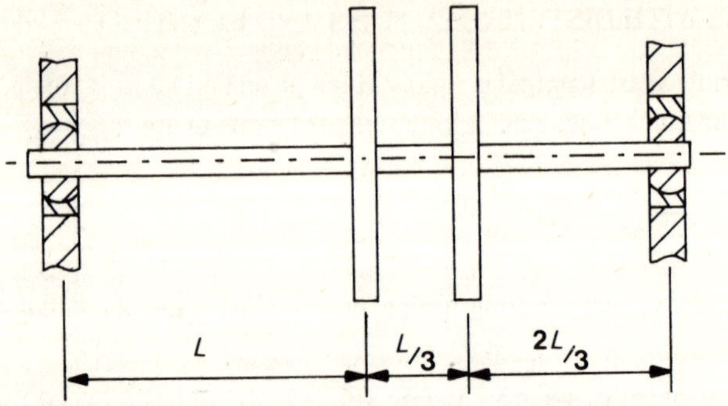

$$\left(\omega^2 = \frac{EI_s/2\,(\pi/2L)^4 L}{7m/4 + \pi^2 I/16L^2}\right)$$

65. Part of the cooling system in a generating station consists of a steel pipe 80 mm in outer diameter, 5 mm thick and 4 m long. The pipe may be assumed to be built in at each end so that the static deflection y at a distance x from one end of the pipe of length L is given by

$$y = \frac{mg}{24EI}\,x^2(L-x)^2,$$

where m is the mass per unit length.
Calculate the lowest natural frequency of transverse vibration of the pipe when full of liquid having a density of 930 kg/m^3. Take the density of steel as 7750 kg/m^3 and E as 200 GN/m^2.

(25.6 Hz)

66. A shaft supported at two positions A and B has a thin disc attached at C, as shown. The bearing at A can be considered long, so that no change in

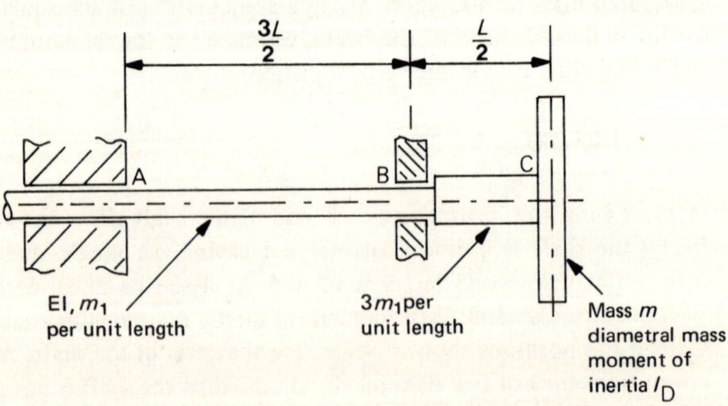

the slope of the shaft at the end can occur, whilst the bearing at B can transmit no bending moment to the shaft. The portion BC of the shaft can be considered as a rigid thin rod.
Estimate the lowest natural frequency of flexural vibrations. You are advised to use a combination of sinusiodal functions for the deflected shape.

$$\left(\omega^2 = \frac{EI/2\,(\pi/L)^4\,3L/4}{[((\pi^2+5)/8)\,m_1 L + \pi^2 m/4 + \pi^2 I_D/L^2]} \right)$$

67. The figure shows a round steel rod, 3 mm in diameter and 300 mm long, fixed rigidly to a firm base at its lower end and having a steel disc 180 mm in diameter and 4 mm thick fixed to its upper end. Young's modulus for steel is 207 GN/m^2 and the density of steel is 7000 kg/m^3; ignore the mass of the rod in comparison with that of the disc, and find (a) the frequency of small flexural vibrations of the system as given, and (b) the smallest mass which, when placed centrally on the disc, would make the system unstable.

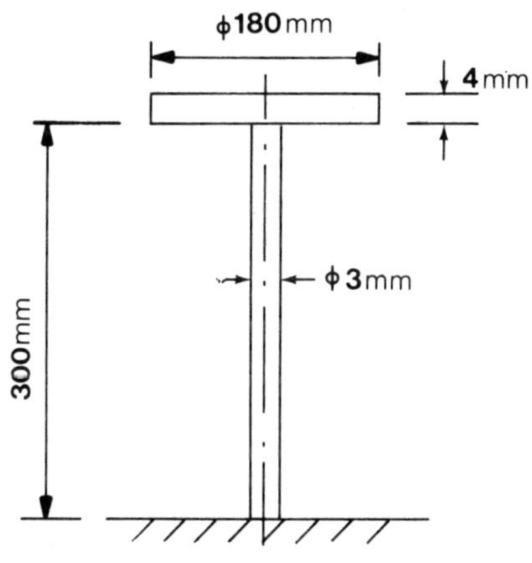

(1,48 H, 1.59 kg)

68. A uniform horizontal beam, made of steel having Young's Modulus equal to 207 GN/m^2, is supported as shown. The beam is rigidly built in to a fixed support at A, and is free to rotate, but not to move vertically, at B. A mass of 560 kg is attached to the beam at C. The relevant flexural second moment of area of the beam is 2.2×10^{-5} m^4, and the beam's own mass and the load on it (excluding the mass at C) are equivalent to a uniformly distributed

mass, moving with the beam, of 600 kg/m. Use Rayleigh's method to estimate the lowest natural frequency of flexural vibrations in the vertical plane.

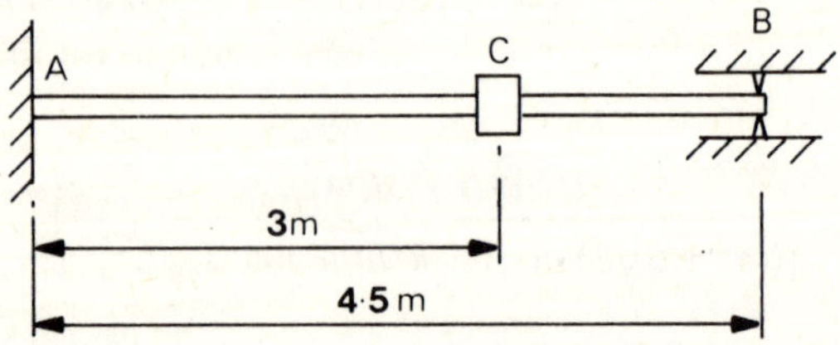

(8.9 Hz)

69. Derive the frequency equation for flexural vibration of a uniform beam which is pinned (simply supported) at one end and free at the other.
Show that the fundamental mode of vibration has a natural frequency of zero and explain the physical significance of this mode.
Obtain an approximate value for the natural frequency of the first bending mode of vibration and compare this with the corresponding value for a beam which is rigidly clamped at one end and free at the other.

70. A rotor of mass 90 kg is fixed at the mid-point of a shaft 1.8 m long. The ends of the shaft are freely supported in self-aligning bearings. If $EI = 10.8$ GN/m^2 for the shaft, and the motion is undamped, calculate the whirling speed.
Damage to the rotor will occur if the amplitude of the axis of rotation of the rotor whilst running exceeds twice its eccentricity. Find the speed range which must be avoided in the absence of other restraints.

6.4 CONTROL SYSTEMS

71. For the hydraulic servo shown in Fig. 5.15 find the value of x_o as a function of time following a sudden change in x_i of magnitude X.

72. A simple position control system contains the following elements:
a device to produce a signal proportional to the difference at any instant between the desired and actual positions;
a linear amplifier; and
a motor which exerts a torque proportional to the input signal to the motor.
The object whose position is to be controlled has a constant inertia I and is supported in a fixed frame by bearings which exert viscous damping of coefficient c.

Draw a block diagram showing how these elements can be arranged to perform the intended function, and show the appropriate transfer operator in each block. Find, in terms of I and c, the value of the factor relating the motor torque to the error signal which will ensure that when a step displacement is required the first overshoot will be one tenth of the magnitude of the step.

($0.718\ c^2/I$)

73. The position of a gun turret, having a constant inertia I and rotating in bearings which give a viscous damping coefficient c, is to be controlled by an electric servo system which consists essentially of
   (a) a potentiometer to give a voltage proportional to the position demanded,
   (b) a second potentiometer to give a voltage proportional to the position achieved at any instant,
   (c) an amplifier which produces an output voltage proportional to the difference at any instant between these two voltages (i.e. the error), and
   (d) a motor which exerts a torque proportional to the voltage applied to it.

   Draw a block diagram (not an electrical circuit diagram), showing how these elements would be arranged to give a working system, and show in each block the appropriate transfer operator.

   During tests it is found that the steady state response to a constant velocity input is acceptable, but that when a step displacement input is applied the first overshoot is larger than desired. Show, by adding to the block diagram and solving the relevant equations, that by adding to the amplifier input a signal proportional to the rate of change of the error, the overshoot to a step displacement input can be reduced without changing the steady state response to a constant velocity input.

74. With reference to a position servo having linear characteristics, state the essential features of (a) output velocity feedback, (b) derivative of error control, and (c) integral of error control. State what kind of error can be eliminated by the use of each method.

   A linear position control system has an amplifier-plus-motor constant of k, an inertia I, and viscous damping (in fixed bearings) c. An integral of error signal is to be added to the error signal before amplification. Find the value of the constant by which the integral signal is multiplied which would make the system unstable.

75. A simple angular position control system contains a device to produce a signal proportional to the difference at any instant between the desired and actual positions, a linear amplifier and a motor which exerts a torque proportional to the input signal to the motor. The load whose position is to be controlled has a constant mass moment of inertia I about its axis of

rotation and is supported in a fixed frame by bearings which exert viscous damping of coefficient c.
Draw a block diagram showing how these elements can be arranged to perform the intended function and show the appropriate transfer function in each block.
In a particular case for which $I = 100$ kg m^2 and $c = 50$ Nm s/rad, when a step input is applied to the system whilst at rest the first overshoot is to be one fifth of the magnitude of the step. Find the value of the factor which relates the motor torque to the error signal.

(30 Nm/rad)

76. A simple linear servomechanism consists of a motor which supplies a torque equal to k times the error (θ) between input (θ_1) and output (θ_2). The motor drives a rotational load of mass moment of inertia I and equivalent viscous damping coefficient (less than critical) c.
Draw a block diagram for the mecahnism and write down the equation of motion.
In a particular mechanism, $I = 80$ kg m^2, $k = 2$ kNm/rad error, and $c = 100$ Nm s/rad. If a step rotation of 10° is applied to the input when the mechanism is at rest, find the time to reach zero error the first time, the time to reach the first overshoot value and the magnitude of the first overshoot.
If, when the mechanism is at rest, the input is suddenly rotated at 10 rad/s, find the steady state error. How could this be altered to 10°?

(0.342 s, 0.633 s, 6.73°, 0.5 rad)

77. A linear servomechanism consists of a proportional controller which supplies a torque equal to K times the error between the input and output positons, and drives a rotational load of mass moment of inertia I and viscous damping coefficient (less than critical) c.
Draw a block diagram for the mechanism and write down its equation of motion.
During tests on a mechanism for which $I = 5$ kg m^2 it was found that

(a) an external torque of 100 Nm applied to the output shaft gave a steady state error of 0.2 rad, and
(b) a constant velocity input of 3 rad/s produced a steady state error of 0.3 rad.

If a step displacement of 10° is applied to the mechanism when it is at rest, find the magnitude of the first overshoot.

(1.62°)

78. A ship's rudder has a moment of inertia about its pivot of 370 kg m^2. When the ship is stationary, rotation of the rudder is resisted by a moment $M_1 = 800D\theta_o$ Nm, where θ_o is the rudder angle in radians measured from the straight-ahead position. The position of the rudder is to be controlled by a simple servo system which applies to the rudder a turning moment $M_2 = k(\theta_i - \theta_o)$ Nm, where k is a constant and θ_i is the rudder angle demanded at any instant.
A step input of 0.1 radian is applied: find the value of k which will make the magnitude of the error $(\theta_i - \theta_o)$ equal to 0.01 radian at the first overshoot. When the ship is moving, hydrodynamic forces on the rudder amount to an additional moment $M_3 = 3500\ \theta_o + 700\ D\theta_o$ Nm: with the value of k found above, and the same step input, find the magnitude of the first overshoot and the steady state error.

(1236 Nm/rad, 0.0114 rad, 0.074 rad)

79. The angular position θ_o of a turntable is controlled by a servomechanism which has acceleration feedback. The turntable has a moment of inertia of 30 kg m^2 and viscous damping of coefficient 20 Nm/rad/s. The motor torque is $10\ (\epsilon + k\ (d^2\theta_o/dt^2))$ Nm, where ϵ is the error, in radians, between the position of the input θ_i and the turntable position θ_o, and $k\ (d^2\theta_o/dt^2)$ is the feedback signal proportional to the acceleration of the turntable.
(i) Draw a block diagram for the system and derive its equation of motion.
(ii) Find the value of k for the damping to be critical.
(iii) If the input to the system is a constant angular velocity, derive an expression for the steady-state error. Determine the value of this error when the input is 2 revolutions per minute.

(2s^2, 24°)

80. The block diagram of a certain position control system is shown. Both the input and output potentiometers have a gain K: the gain of the amplifier

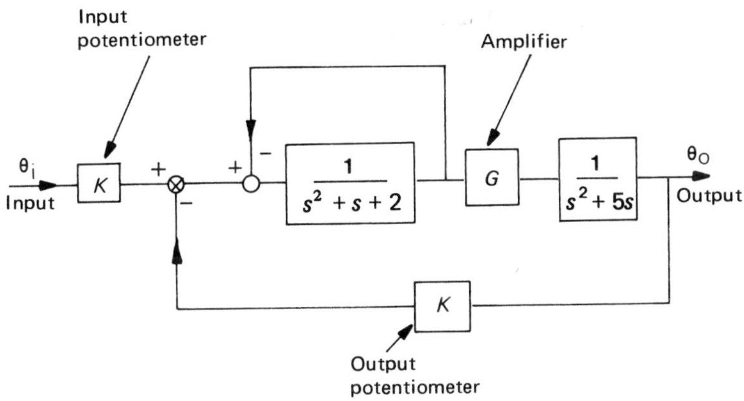

is G. Simplify the system to one having a single unity feedback loop and find the open loop transfer function.
By sketching the root locus diagram for this system determine the maximum value of G for stability.

$(0 < GK < 13.75)$

81. The main elements of a position control system are shown in the figure in block diagram form. The input and output potentiometers each have a gain K, and the gain of the amplifier is G.

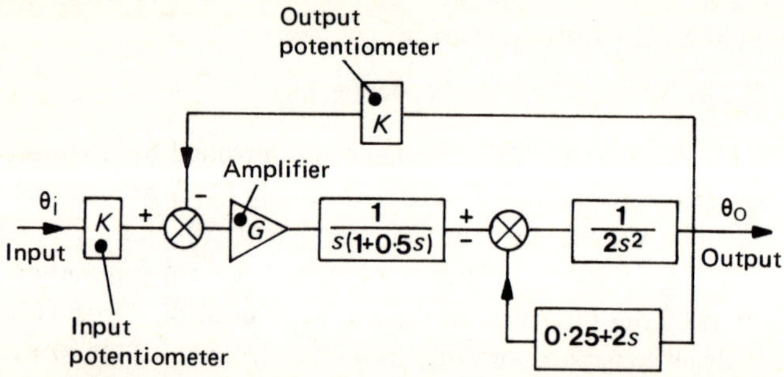

Find the open loop transfer function of the equivalent single loop system, and sketch the root locus diagram for this system. Hence determine the limiting value of the product GK for stability.

(0.17)

82. For the control system shown in the figure in block diagram form, find the OLTF of the equivalent single loop system and sketch the Root Locus diagram for this system if $0 \leqslant K_1 \leqslant \infty$ and $K_2 = 17/32$. Hence find the maximum value of the amplifier gain K_1 for stability.
Also determine K_1 and K_2 so that the system has closed loop poles at $s = -2 \pm j2$.

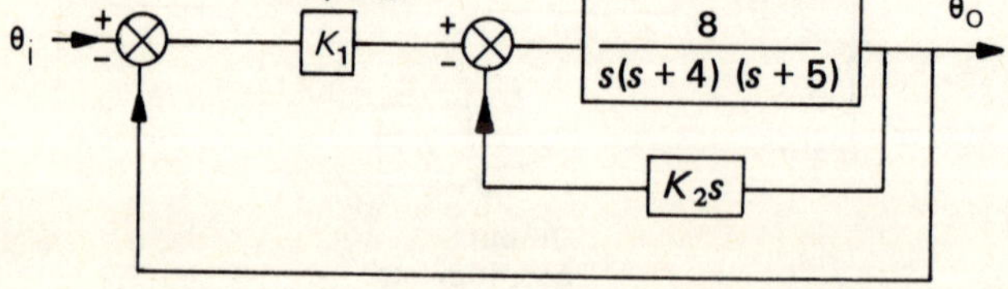

83. A root locus diagram is often used to assess the dynamic behaviour of a system. What does each line or locus on such a diagram represent, and what aspects of the dynamic behaviour does it demonstrate?

84. What is meant by the *stability* of a dynamic system, and how does an unstable system respond to a stimulus?

85. The simple unity feedback control system shown has three elements with individual functions as shown.

(a) Sketch the root locus diagram of this system and determine the maximum value of the amplifier gain K for stability.
(b) Find a value of K such that the closed loop system has a complex conjugate pole pair with damping factor 0.5.

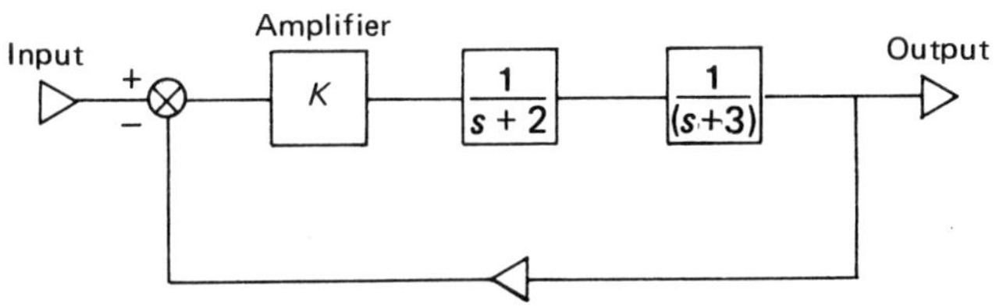

(30, 72)

86. Given the open loop transfer function of a dynamic system, explain how you would construct a Nyquist frequency response diagram. How would you determine from this diagram whether the system was stable or not?

87. A control system has an open-loop transfer function given by

$$\Phi_o = \frac{K}{(s+1)(s^2+5s+6)}$$

where K is an unspecified gain constant.

(a) Derive an expression for the overall or closed-loop transfer function, Φ_c,
(b) For the particular case of $K = 90$, sketch the Nyquist frequency response diagram for the open-loop transfer function of this system and thus determine whether or not the system is stable.
(c) Hence, or otherwise, determine the range of values of K for which the system *is* stable.

$(0 < K < 60)$

88. A control system has an open loop transfer function (OLTF) given by

$$\Phi(s) = \frac{k}{s^3 + 6s^2 + 8s}$$

where k is a gain constant.
Sketch the Nyquist frequency response diagram for the OLTF of this system when $k = 20$ and thus determine whether or not the system is stable.
Also, find the maximum value which k can have for the system to be stable.

(48)

89. A control system has an open-loop transfer function (OLTF) given by

$$\Phi(s) = \frac{2}{s(1 + s)(1 + 2s)}$$

Sketch the Nyquist frequency response diagram for the OLTF of this system and thus determine whether the system is stable or not.
The system is modified by adding to the open loop derivative action with a time constant of 0.5 s.
Write down the OLTF of the modified system and hence sketch the Nyquist frequency response diagram. Comment on the stability of the modified system and calculate the gain margin.

(3)

ACKNOWLEDGEMENT

Some of the exercises first appeared in the following University of London BSc (Engineering) Degree Examinations, set for students at Imperial College of Science and Technology.

Part 2. Mechanics.

1971 (Q 3-7), 1972 (Q 4-8), 1973 and 1974 (Q 3-8),
1975 and 1976 (Q 2-5, 7, 8), 1977 (Q 3-7), 1978 (Q 3-6, 8),
1979 (Q 3-8) and 1980 (Q 2, 3, 8).

Part 3. Vibrations and Control.

1975-1980 (Q 1-3).

Index

V

W

Z